Distributed Systems

Series Editor
Jean-Charles Pomerol

Distributed Systems

Concurrency and Consistency

Matthieu Perrin

First published 2017 in Great Britain and the United States by ISTE Press Ltd and Elsevier Ltd

ISTE Press Ltd
27-37 St George's Road
London SW19 4EU
UK

www.iste.co.uk

Elsevier Ltd
The Boulevard, Langford Lane
Kidlington, Oxford, OX5 1GB
UK

www.elsevier.com

British Library Cataloguing-in-Publication Data
A CIP record for this book is available from the British Library
Library of Congress Cataloging in Publication Data
A catalog record for this book is available from the Library of Congress
ISBN 978-1-78548-226-7

Printed and bound in the UK and US

Contents

Acknowledgments

I am grateful to Achour Mostéfaoui and Claude Jard for their contribution to the work presented in this book. I offer warm thanks to Luc Bougé and Maria Potop-Butucaru for their relevant comments that helped improve the quality of this book. Finally, I would like to thank Matoula Petrolia for playing Alice's role in the illustrating examples.

Introduction

I.1. Preamble

Bob was daydreaming as he walked into the subway station. It has been two days since he had last talked to Alice. His anger has gradually turned into bitterness, then into guilt. He understood now that his initial reaction had been somewhat exaggerated and wanted more than anything to bury the hatchet with his childhood friend but did not know how to approach the situation. He felt surprised and relieved when he received a message from her inviting him to go out for coffee. He managed to quickly respond as the train was approaching the dark tunnel. However, at the heart of the city, the failing network could not transmit his liberating reply “Obviously”.

Alice loved Bob as a sister, despite the chronic anxiety and shady character of the one whom she saw as her oldest friend. Two days earlier, he had left in a rage because of a simple misunderstanding and she hadn’t been able to reason with him since. Even now, reconciliation seemed difficult: Bob had purely and simply ignored her invitation to go out for coffee. She tried a new strategy and sent “You didn’t answer.” followed a few seconds later by a humble “Are you angry?”.

The story does not tell what happened in the end and we can only imagine it. As strange as it may seem, Alice and Bob’s fate very strongly depends on the instant messaging service chosen to communicate. We have tried to reproduce their conversation with three general instant messaging services: Google Hangouts[1] (Hangouts), WhatsApp Messenger[2] (WhatsApp) and Microsoft Skype[3] (Skype). The experiment has been performed with Matoula Petrolia playing the role of Alice and myself in the role of Bob. Bob’s dropping connection has been modeled by using the

1 http://www.hangouts.google.com/

2 http://www.whatsapp.com/

3 http://www.skype.com/

airplane mode of the mobile phone. For each experiment, we present a screenshot taken before reconnection and another taken after reconnection for both interlocutors. The three messaging services exhibit very different behaviors confronted with the same situation.

Hangouts (Figure I.1). Bob received Alice's messages when coming out of the subway. He looked absently at them thinking to himself that his own message should have taken some time to reach her. It was not until late in the afternoon, when he was about to specify the time and place of the meeting that he noticed the error message written in small prints underneath her answer: "Message not sent. Touch to retry.". He resigned to call Alice to straighten things out.

WhatsApp (Figure I.2). Bob's response felt to Alice like a cold shower. "Obviously!" Not only he was angry at her but he has had no problems in saying it out loud and as sarcastically as possible. Bob asked her later at what time she wanted to see him. She did not immediately understand but seized the opportunity to accept. It was only in the evening, when Bob showed her his own thread of messages that she understood what had happened: their messages had been mixed up and each had received the message from the other only after having sent their own. As a result, Bob couldn't have known that she had misinterpreted his message.

Skype (Figure I.3). Alice's second strategy was also as unsuccessful as the first. Although she was regularly checking her phone the whole day long, there was no message from Bob arriving after her own on her thread of messages. She felt stupid when she finally realized her mistake: she had not seen the cheerful response that Bob had sent her before she sent him her second message. She hastened to apologize and to make another meeting with Bob.

Based on these three scenarios, it can be seen that compromises are inevitable when temporary disconnections happen, be it an error message (Hangouts), a reordering of messages (Skype) or a presentation of a different state to the interlocutors (WhatsApp). Each strategy has its own qualities but none is without flaw. Instant messaging services are intended to be used by human beings, capable of adapting themselves to numerous situations. Programs, for their part, are much less able to do so and the consequences of unexpected inconsistencies in the data shared by computers can be much more serious. Disposing of an effective way to precisely model the different types of inconsistency that can occur is therefore a very important challenge.

I.2. Distributed systems and concurrency

DEFINITION I.1.– A *distributed system* is a collection of autonomous computing entities connected to accomplish a joint task.

Detailed below are the three elements important to this definition.

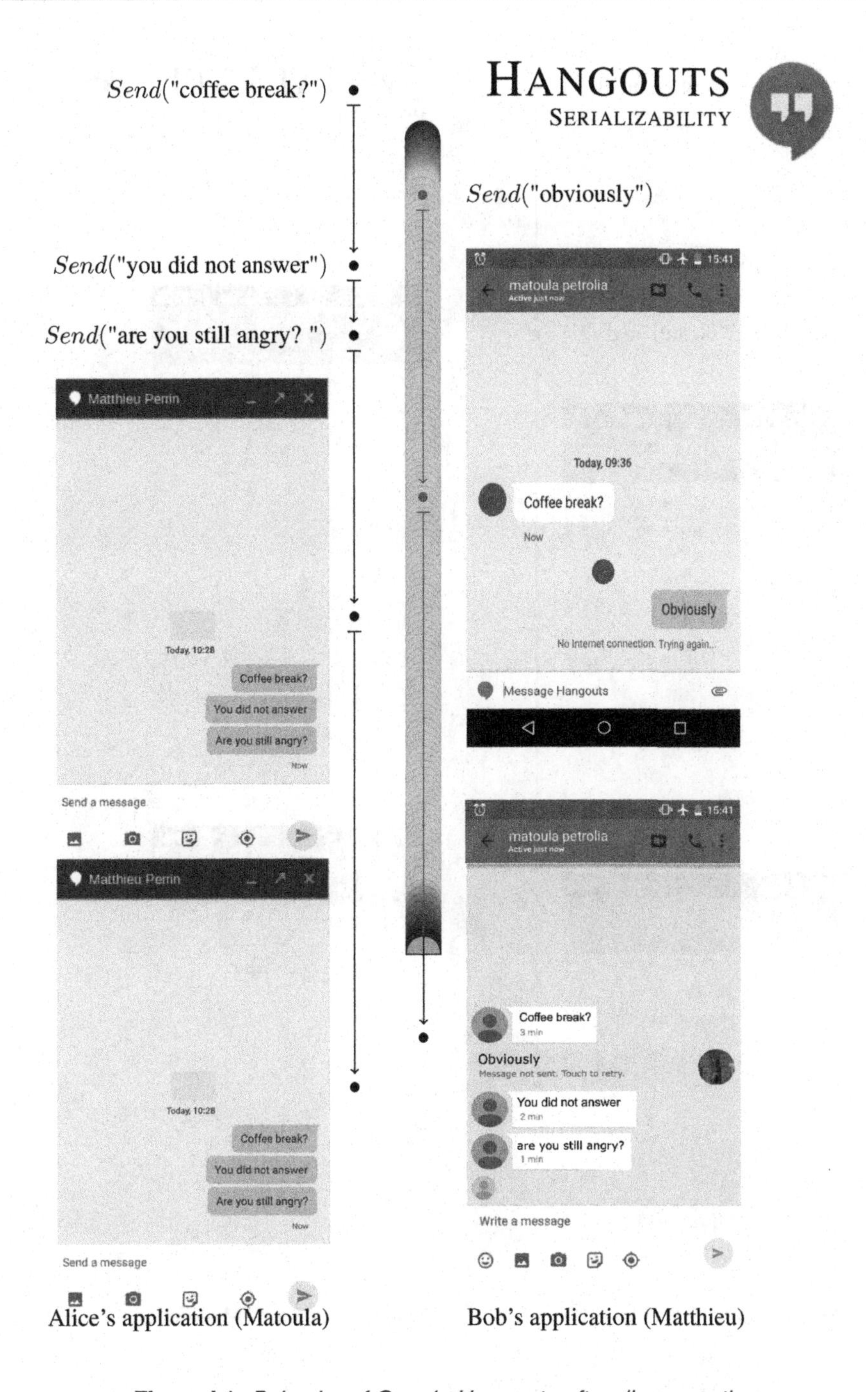

Figure I.1. *Behavior of Google Hangouts after disconnection*

WHATSAPP
PIPELINED CONSISTENCY

$Send$("coffee break?")

$Send$("obviously")

$Send$("you did not answer")

$Send$("are you still angry?")

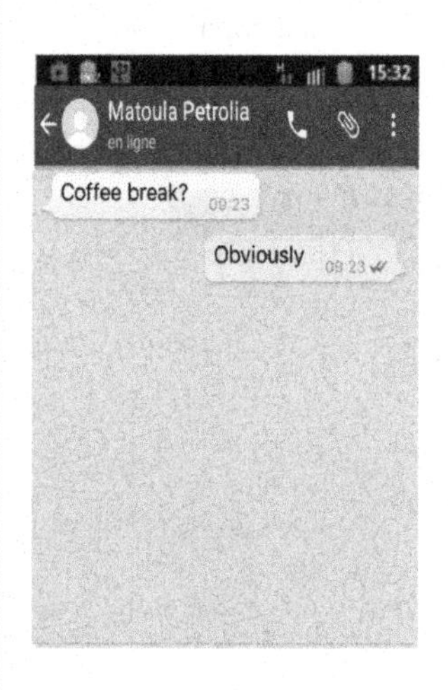

matthieu perrin
online
Coffee break?
You did not answer
Are you still angry?

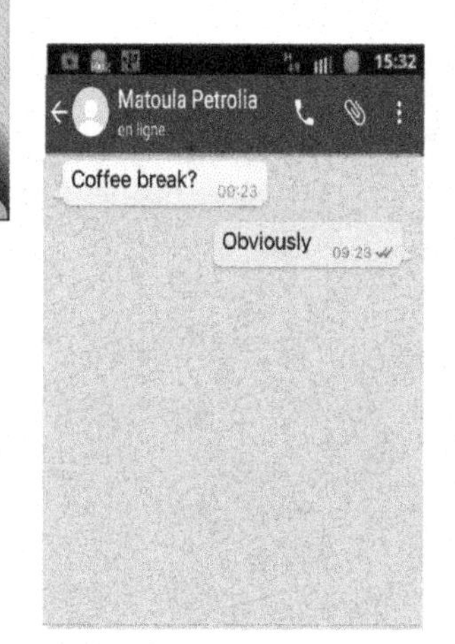

matthieu perrin
online
Coffee break?
You did not answer
Are you still angry?
Obviously

Alice's application (Matoula) Bob's application (Matthieu)

Figure I.2. *Behavior of WhatsApp Messenger after disconnection*

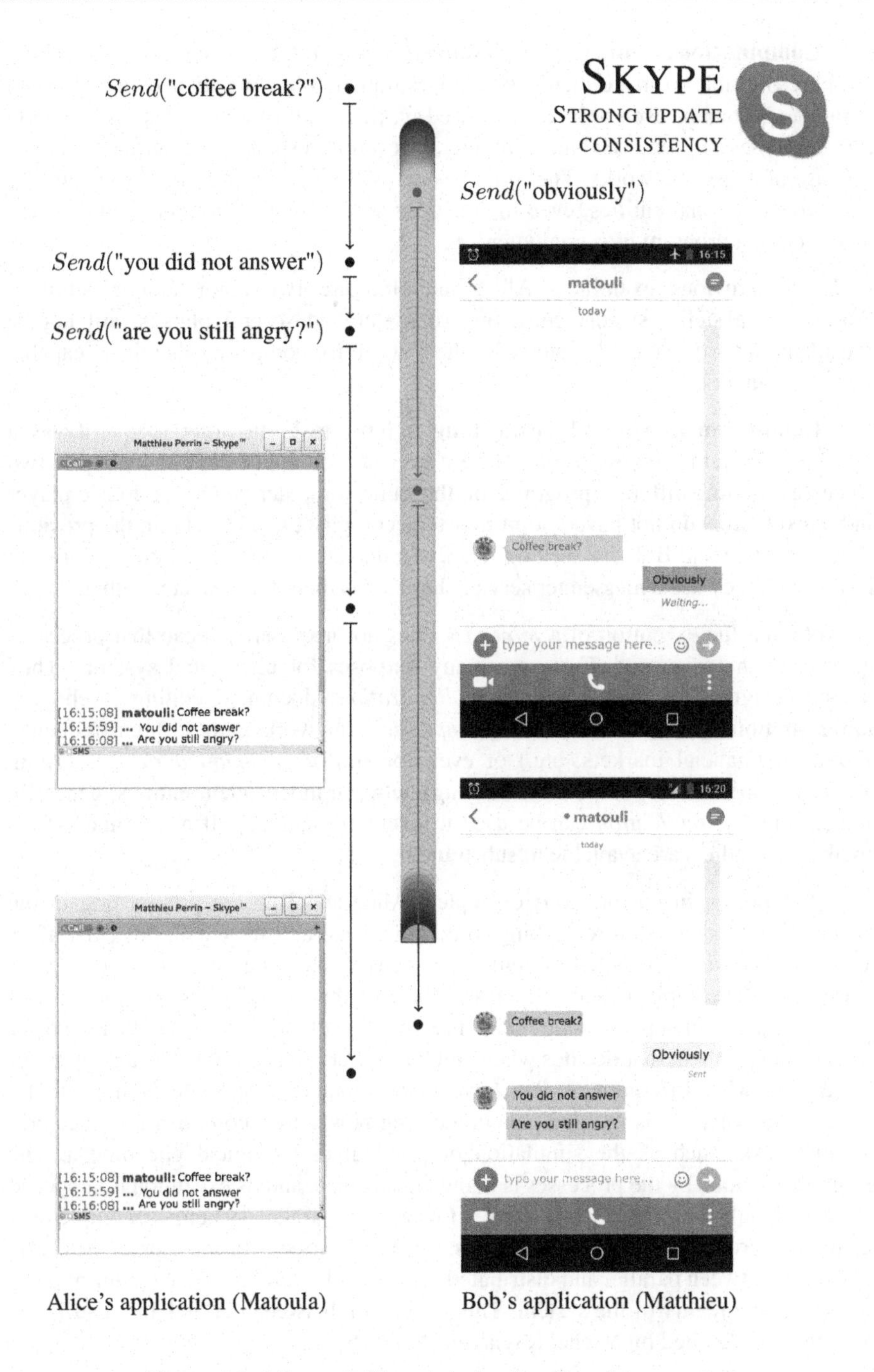

Figure I.3. *Microsoft Skype's behavior after disconnection*

Computational entity. *Computational entity* refers to entities sufficiently complex and able to make their own choices, either following their own will, such as human beings or even animals, or based on the result of a complex computation when it comes to processes on a computer, or even in response to external stimuli in the case of sensor networks. The term *process* will be generically used to designate these computational entities, even in situations where they do not correspond to the execution of a program by a computer.

In the previous examples, Alice and Bob are two computational entities. Conversely, a stellar system consisting of a star and several planets will not be considered as a distributed system, because no entity comprised therein is capable of decision-making.

Connection in view of performing a joint task. In order to qualify as a distributed system, processes must have a need to communicate. For example, two processes running different programs on the same computer (such as a music player and a text editor) do not have a joint task to accomplish. Conversely, in the previous example, Alice and Bob are seeking to solve a dispute, which forms their joint task. To this end, they use a messenger service that allows them to connect together.

Note that the execution of a *joint task* does not necessarily mean that processes share a common interest. There are many purposes for distributed systems. They can be designed for the purpose of *collaboration* (document editing, Web sites administration, etc.) or instead for *competition* (network games, high-frequency trading in financial markets, etc.) or even for simple *communication* (messaging services, websites, peer-to-peer file sharing) between independent entities, generally human. Finally, *replication* can be used to overcome failures: if a machine suffers break down, others are available to substitute it.

Autonomy. In the previous examples, Alice and Bob are pre-existing in the system. They existed before seeking to communicate and they will still exist after. In the analysis of distributed systems, the focus is on issues that may occur with processes to accomplish their joint task. The fact that they are autonomous means that their operation depends only on themselves. In particular, there can be no master process among them that decides when and how others can operate. The autonomy of processes is what differentiates *distributed* systems and *parallel* systems. In a parallel system, the objective is to aggregate computing power to accomplish a particularly complex task, such as the simulation of physical or biological phenomena. The program executed by the processes is known in advance, and its execution is supposed to depend only on initial parameters. However, a parallel system can implement techniques proper to distributed systems, especially to synchronize processes. The difference between parallel and distributed systems is therefore more a matter of point of view than a reality of the system. This distinction between parallel and distributed computing is detailed by Michel Raynal in [RAY 15].

The world that surrounds us is inherently a distributed system composed of independent individuals forced to communicate in their daily tasks to elect their representatives, to avoid collisions on the road, etc. Specifically, distributed *computer* systems, in which the behavior of processes is governed by controllable programs, have recently invaded our lives. In 2015, over three billion internauts were recorded throughout the world[4], today any processor in a mobile phone is multi-core, cloud computing is starting to become more democratic with the expansion of the services it offers and geo-replication is widely used to ensure the persistence of sensitive data.

Despite all possible applications of distributed systems, distributed applications are generally much harder to design than their sequential counterparts: while events in a sequential environment occur one after another in a sequence controlled by the programmer, the notion of time between events occurring in a distributed system appears to be much more blurred. In Alice and Bob's example, it is not clear if Alice has sent her last two messages *before* or *after* Bob, hence the differences in behaviors observed in this situation for the three instant messaging services. This situation, proper to distributed systems, in which events are not totally ordered has a name: concurrency.

DEFINITION I.2.– A system is said to be *concurrent* if its behavior depends on the order in which operate the actors of the system.

I.3. Wait-free distributed systems

The mastering of concurrency is one of the main challenges in distributed systems. According to the security level required for applications and the control of incidents likely to occur in the system in which they are deployed, concurrency will be more or less visible. For example, in the control system of a plane, the relative speed of each component and the paths taken by control orders are perfectly under control so as to remove any uncertainty associated with concurrency. In contrast, in collaborative document editors such as subversion or git, the majority of editing events are carried out offline. Concurrency is then detected at the time of synchronization, resulting in conflicts that must be manually solved. The main parameters that characterize a distributed system are listed here below.

System scale. A system is distributed when it comprises at least two processes. Therefore, a program whose graphical interface (GUI) is displayed on the screen in an asynchronous fashion can already be considered as distributed. In contrast, Tianhe-2, currently the most powerful supercomputer in the world, has 3,120,000 cores and BOINC, the management platform of the peer-to-peer project SETI@home [SPA 99] dedicated to the research of extra-terrestrial signals, claims to be installed on more than thirteen million computers in the world[5].

---

4 http://www.internetlivestats.com/watch/internet-users/

5 http://boincstats.com

Interaction means. In general, we can make a distinction between systems in which processes must communicate by exchanging messages and those in which processes access shared memory. Finer divisions can bring further detail into these families. If processes communicate by messages, is the network complete? Do processes know their neighbors? Are these fixed? Can they choose their neighbors? For those communicating by means of shared memory, what operations are they allowed to call? Can they write to all registers?

Failure management. The greater the systems, the more frequent failures are. In peer-to-peer systems, processes leave and others connect to one another on a permanent basis; when one considers the Internet as being increasingly composed of connected mobile terminals, the reliability of transmissions cannot be assured; hackers will always seek to attack sensitive applications such as those pertaining to banking and governmental institutions (the so-called Byzantine faults). These faults can be of various kinds: crash failure (a process stops its execution without warning others), omission failure or loss of messages (when sending or receiving) or non-compliant process behavior (memory corruption or intentional attacks). The number of failures that can be accepted and the presence of a reliable server known to all are also defining the characteristics of the system.

Relation to time. The presence or lack of a global clock accessible by all processes, completely changes the nature of a system. In a fully synchronous system, all processes proceed according to a well-defined temporal pattern (eventually *a priori*) and if they communicate through messages, these messages are transmitted within a limited timeframe. These two assumptions can be impaired. In a completely asynchronous system, there is no bound on the relative speed of processes or on the duration of message transfers (network *latency*). If, in addition, crash failures can occur, a process that does not receive an expected message is unable to deduce if the transmitter is faulty or extremely slow. The heterogeneity of the hardware running the processes can be a major cause of asynchronicity.

The purpose of this book is to study concurrency. We are therefore considering a poorly performing system where applications cannot hide it completely. For example, it could be accepted, as in the preface, that processes undergo temporary disconnections. More specifically, we assume that processes communicate by sending messages to one another but that it is not acceptable nor possible for a process to wait explicitly for messages from an another process. This assumption is realistic to model a large number of systems, including those mentioned below.

Size. Systems whose number of processes is unknown, which prevent knowing the number of responses to expect, in peer-to-peer systems, for instance.

Cost. Systems for which time spent waiting is considered too costly, for example in high-performance computing.

Partitions. Systems in which partitions can occur, temporarily preventing processes to communicate with each other. Such partitions are frequent in cloud architectures [VOG 09]. Temporary disconnections on the Internet can also be modeled by partitions.

Faults. Systems for which fault management is critical to the extent where the system should continue to operate even if only a single process is left operating in the system. In such a situation, a process cannot wait for the participation of a known *a priori* number of other processes as they may be at fault. This case is the one that we prefer to model the hypothesis in which there is no waiting time because it is easier to formally define.

More precisely, we consider wait-free message-passing asynchronous distributed systems. We only provide the intuition here, these systems being formally defined in section 1.3.2.

Distributed message-passing systems. The systems that we consider are composed of a fixed and known-sized set of sequential processes that communicate by means of primitives for sending and receiving messages.

Wait-free systems. All processes are likely to crash and the system must continue to operate even when reduced to a single process.

Asynchronous systems. Processes do not run at the same speed and there is no bound on the time taken by a message to reach its destination. Under these conditions, a process that does not receive a message that it is expecting can never know if this message is very slow to reach it, or if the message has never been sent because of failure in the transmitter.

I.4. Shared objects

At a very high level, distributed applications all share a similar model: *processes* interact with each other by means of one or more shared objects that are able to manage concurrency. In the case of an online game, the shared object is the game being played and the rules of the game make explicit authorized strokes; processes that compute weather forecasts share the map during the simulation, etc. Shared memory and communication channels are other examples of shared objects, but at a lower level. In general, distributed applications can often be modeled by layers, each layer being an abstraction in the form of shared objects of the previous layer. To sum up, a program can be seen as a hierarchy of shared objects, each with its level of semantic requirement.

For instance, consider a collaborative editing of documents. At the highest level, the application provides access to character insert and delete operations at the cursor level and cursor movement operations, accessible from the keyboard. The implementation of these operations involves a data structure that stores the sequence of characters of the document. Depending on the application, this sequence itself utilizes lower-level objects, such as a data management system abstracting the communication primitives of the system in which the application is implemented.

The fact of considering a perspective that involves wait-free systems has significant implications for the shared objects that can be build therein [ATT 95, ATT 94]. When an operation on a shared object is initiated at the local level by a process, it must be processed using only local knowledge of the process. The reads should be effected locally, which means that each process has a local copy of the object. The impossibility of synchronizing during operations prevents the removal of all inconsistencies. These must therefore be specified with special attention. The central question of this book can be formulated in this way: *how can shared objects be specified in a wait-free system?*

I.5. Shared object specification

How can we describe what an instant messaging service is? If we ask the question randomly to users of these services, odds are high that the answer will be as follows. "An instant messaging service allows the sending of messages between multiple correspondents and displays received messages, from the oldest to the newest". This describes the *normal* behavior of these services, which occurs when there is no concurrency occurring, i.e. in the case of sequential use. In the same way, developers of concurrent applications in general often wish they could make use of a shared version of the objects they are accustomed to in sequential programs. The specification of a shared object must therefore rely on its *sequential specification.*

Obviously, the sequential specification is insufficient to explain the differences between the three services exposed in the prologue. One additional property, *consistency criterion*, is necessary to explain how concurrency is managed. During concurrent execution, accesses to shared objects create interactions between the local operations of the various processes. Figure I.4 shows some examples of histories extracted from concurrent executions involving one or two processes sharing a set inside which it is possible to insert values or whose content can be read, i.e. the set of the values previously inserted. The figure shows the operations performed by each process over time, which elapses from the left to the right. A rounded rectangle defines the time between the call and the return of an operation. The values returned by the reads are represented after a slash.

Which histories are correct? In the sequential case, there is no possible ambiguity. In Figure 1.4(a), during its first read, the process has inserted 1, it can therefore return only $\{1\}$. The reads following the insertion of the 2 have no other option than returning $\{1, 2\}$. The history of Figure 1.4(b) must also be considered as correct: all operations are consecutive in time and the sequence of values returned is the same as in the sequential execution. What about Figure 1.4(c)? The operations are also sequential but the first read follows the insertion of the 2 without taking it into account. On the one hand, we could argue that this reading is false and therefore that this history is inconsistent. On the other hand, each process performs the same operations and reads the same values as in the previous execution. The only difference between these two histories is thus how both processes perceive their relative speed [PER 16]. However, in an asynchronous system, processes are unable to tell the difference between the two histories, because the observed temporal order could also be due to a desynchronization of the clocks of both processes, which would have no effect on programs that use the object. There is therefore no definitive answer to what is a consistent history, every situation being simply better described by a particular consistency criterion. History of Figure I.4(b) is said to be *linearizable* whereas that of Figure I.4(c) is only *sequentially consistent*.

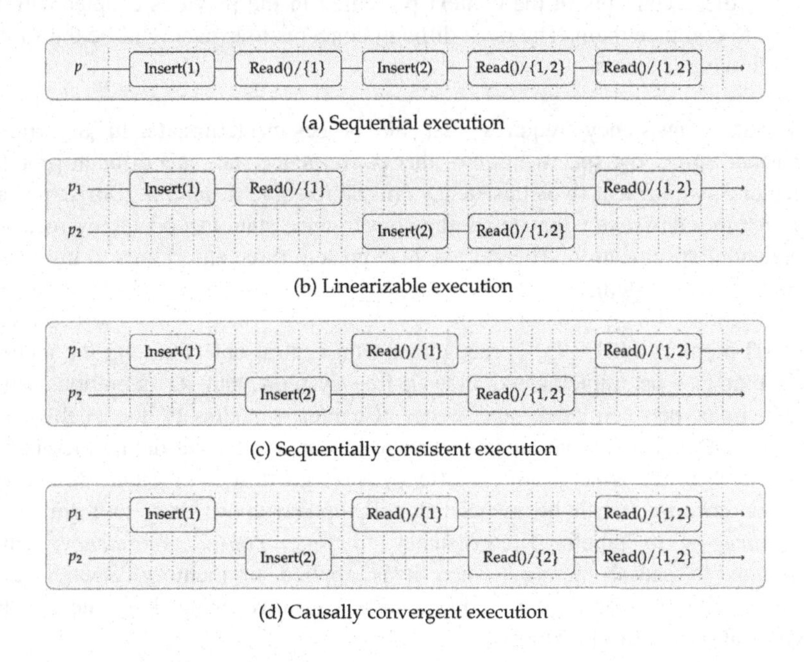

Figure I.4. *Which behaviors are acceptable?*

Unfortunately, sequential consistency and linearizability are too strong criteria to be implemented in wait-free systems [ATT 94, GIL 02]: inconsistencies such as those of history in Figure 4(d) cannot be avoided. In this case, how can we describe such histories? The literature related to the subject presents a limited picture of the concepts and definitions, often partial and incomparable between themselves. The extension and the formalization of these concepts is an important part of the work presented in this book. We endeavor to fill in the blanks in order to extract the structure of the space of weak criteria and therefrom to draw a map presented in Figure I.5.

I.6. Organization

After this introduction, a notation table is provided. Next, Chapter 1 presents the mathematical framework common to all other chapters: therein we formally define sequential specifications, concurrent histories and consistency criteria. This chapter also formally defines wait-free systems.

In Chapter 2, we present the key concepts proposed by the various communities facing concurrency to specify their distributed objects. As much as possible, we try to include these concepts in the context presented in the previous chapter with the objective to compare them. The most difficult concepts to expand will be the subject to the following chapters.

Eventual consistency requires that all processes terminate in a common *convergence state*. The specification of this convergence state is a difficult problem. In Chapter 3, we present two consistency criteria to solve it: *update consistency* and *strong update consistency* that relate the convergence state to updates by means of the sequential specification. Therein, we also present three algorithms to implement them in wait-free systems.

Causal memory [AHA 95] is recognized as a central object among the memory models that can be implemented in wait-free systems, but its definition utilizes semantic links between reads and writes. Chapter 4 addresses the challenge of including causality as a consistency criterion in our model. We define therein four consistency criteria. *Weak causal consistency* is the greatest common denominator of these four criteria. It can be associated with update consistency to form *causal convergence* or to pipelined consistency to form *causal consistency*, which corresponds to causal memory when it is applied to memory. *Strong causal consistency* finally enables the study of false causality induced by the standard implementation of causal memory.

Chapter 5 focuses on computability in wait-free systems. In this chapter, we establish a structural map of the space of weak criteria, divided into three families of primary criteria (*eventual consistency*, *validity* and *state locality*) which can be combined in pairs to form three families of secondary criteria (*update consistency*, *pipelined consistency* and *serializability*). On the other hand, the conjunction of the three primary criteria cannot be implemented in wait-free systems, which justifies the colors of Figure I.5.

Chapter 6 shows how these concepts can be applied to programming for distributed systems. The library *CODS* implements weak consistency criteria within the object-oriented programming language D. CODS offers the most possible transparent abstract programming interface, in which instantiation is only replaced by the reference to a sequential specification defined by a class of the program and of a consistency criterion among those of the library or even implemented as an extension of this library.

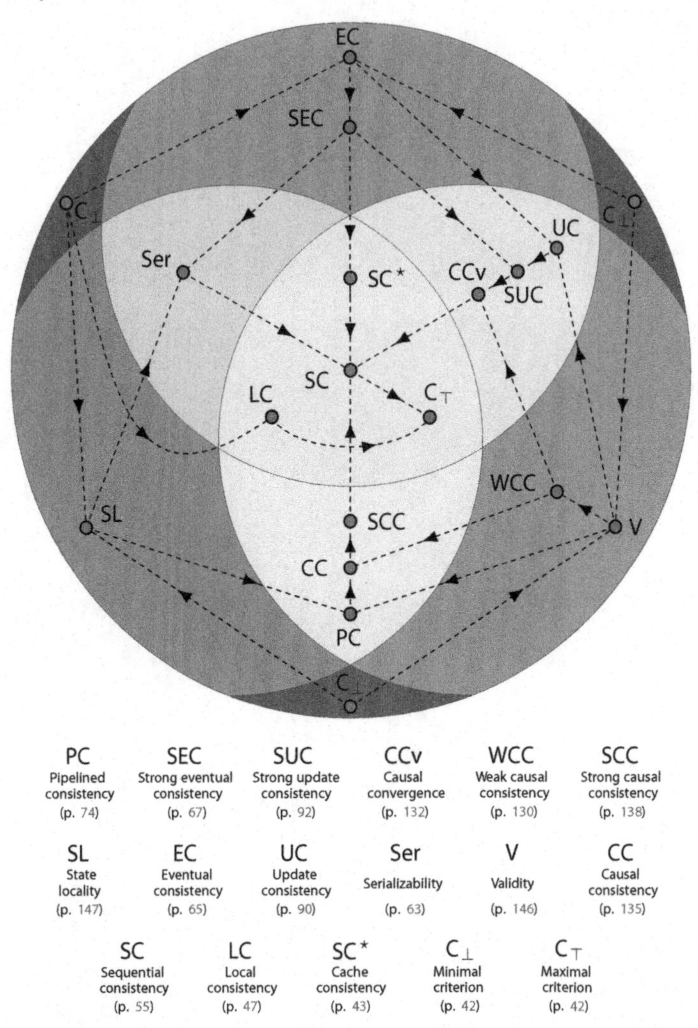

Figure I.5. *Map of the consistency criteria discussed in this book. For the color version of this figure, www.iste.co.uk/perrin/distributed.zip*

Finally, Chapter 6 summarizes the previous chapters and outlines future directions for research. After the bibliography, the index provides help to navigate the document.

List of Notations

Mathematical notations

$f: \begin{cases} E \to F \\ x \mapsto f(x) \end{cases}$	Function f that maps $f(x) \in F$ to each $x \in E$
$\mathcal{P}(E)$	Power set of E
$\mathcal{P}_{<\infty}(E)$	Set of subsets of finite cardinality of E
$E \sqcup F$	Disjoint union of E and F
$f\|_E$	Restriction of f to E (f: function, E: set)
$\lfloor e \rfloor_{\prec}$	Predecessors of e in $\prec$ ($\prec \subset E^2$: acyclic relation, $e \in E$)
$\lceil e \rceil_{\prec}$	Successors of e in $\prec$ ($\prec \subset E^2$: acyclic relation, $e \in E$)
$\prec^\star$	Transitive closure of $\prec$ ($\prec \subset E^2$: relation)
$\Sigma^\star$	Set of finite words on Σ (Σ: alphabet)
Σ^ω	Set of infinite words on Σ (Σ: alphabet)
Σ^∞	$\Sigma^\infty = \Sigma^\star \cup \Sigma^\omega$ (Σ: alphabet)
$\|u\|$	Length of u ($u \in \Sigma^\star$)
ε	Word of length 0
$u \cdot v$	Concatenation of u and v ($u \in \Sigma^\star$, $v \in \Sigma^\infty$)
$[a - z]$	Set of the letters in the Latin alphabet

Abstract data types

ADT	Abstract data type
UQ-ADT	Update-query abstract data type
CADT	Commutative abstract data type
$\mathcal{T}$	Set of all ADTs
$\mathcal{T}_{UQ}$	Set of all UQ-ADTs
A (read *Alpha*)	Input alphabet

B (read *Beta*)	Output alphabet
Z (read *Zeta*)	Set of abstract states
ζ_0	Initial state
τ	Transition function
δ	Output function
Σ	Set of operations
τ_T	Transition function for the operations ($T \in \mathcal{T}$)
δ_T^{-1}	Possible states for the operations ($T \in \mathcal{T}$)
U_T	Set of update operations of T ($T \in \mathcal{T}$)
Q_T	Set of query operations of T ($T \in \mathcal{T}$)
$\hat{U}_T$	Set of pure update operations of T ($T \in \mathcal{T}$)
$\hat{Q}_T$	Set of pure query operations of T ($T \in \mathcal{T}$)
$L(T)$	Sequential specification of T ($T \in \mathcal{T}$)
$T \times T'$	Composition of T and T' ($T, T' \in \mathcal{T}$)
$\mathcal{M}_x$	Register x (x: register name)
$\mathcal{M}_X$	Memory on X (X: set of register names)
$\mathcal{S}_{Val}$	Set of support Val (Val: set of values)
$\mathcal{Q}_X$	Queue ADT on X (X: set of values)
$\mathcal{Q}'_X$	Queue UQ-ADT on X (X: set of values)
$\mathcal{W}_k$	Sliding window register of size k ($k \in \mathbb{N}$)

Concurrent histories

$\mathcal{H}$	Concurrent histories
$\text{lin}(H)$	Linearizations of H ($H \in \mathcal{H}$)
E_H	Events of H ($H \in \mathcal{H}$)
$U_{T,H}$	Update events of H ($T \in \mathcal{T}, H \in \mathcal{H}$)
$Q_{T,H}$	Query events of H ($T \in \mathcal{T}, H \in \mathcal{H}$)
$\hat{U}_{T,H}$	Pure update events of H ($T \in \mathcal{T}, H \in \mathcal{H}$)
$\hat{Q}_{T,H}$	Pure query events of H ($T \in \mathcal{T}, H \in \mathcal{H}$)
$e \mapsto e'$	Process order ($e, e' \in E_H$)
$e \xrightarrow{\text{VIS}} e'$	Visibility relation ($e, e' \in E_H$)
$Vis(H)$	Set of visibility relations of H ($H \in \mathcal{H}$)
$e \leq e'$	Total order ($e, e' \in E_H$)
$e \dashrightarrow e'$	Causal order ($e, e' \in E_H$)
$co(H)$	Set of causal orders ($H \in \mathcal{H}$)
Σ	Set of operations
Λ	Labeling function
$H^{\dashrightarrow}[\mathsf{A}/\mathsf{B}]$	Projection ($H \in \mathcal{H}, \dashrightarrow \in (E_H)^2, \mathsf{A}, \mathsf{B} \subset E_H$)
$\mathscr{P}_H$	Maximal chains of H ($H \in \mathcal{H}$)
$AS_n[\varphi]$	Asynchronous message-passing system (n: number of processes, φ: condition on the faults)

Consistency criteria

$\mathcal{C}$	Set of consistency criteria
$\mathcal{C}_W$	Set of weak criteria
$\mathcal{C}_S$	Set of strong criteria
$C_1 \leq C_2$	C_2 is stronger than C_1 ($C_1, C_2 \in \mathcal{C}$)
$C_1 + C_2$	Upper bound of C_1 and C_2 ($C_1, C_2 \in \mathcal{C}$)
$C_1 \Delta C_2$	Lower bound of C_1 and C_2 ($C_1, C_2 \in \mathcal{C}$)
$C(T, T')$	Composition of shared objects ($C \in \mathcal{C}, T, T' \in \mathcal{T}$)
$C_\top$	Maximal criterion
$C_\bot$	Minimal criterion
$C^\star$	Composition closure of C ($C \in \mathcal{C}$)
SC	Sequential consistency
PC	Pipelined consistency
LC	Local consistency
Ser	Serializability
V	Validity
SL	State locality
EC	Eventual consistency
SEC	Strong eventual consistency
UC	Update consistency
SUC	Strong update consistency
WCC	Weak causal consistency
CC	Causal consistency
SCC	Strong causal consistency
CCv	Causal convergence

1

Specification of Shared Objects

1.1. Introduction

When a project manager designs the general architecture of an application, one of his main concerns is to divide the workload into distinct logical entities, software building blocks, thus enabling them to be implemented by different developers. Each software building block is then separately specified to mitigate the risks of conflict during the final integration. The specification is thus the communication vector between two developers: the *designer* who designs the object and the *user* who uses it in his/her own program. When the object in question is part of a library (and all the more when the source code is proprietary), this is the only way to describe the object. A good specification should be written from the user's perspective and for the user. We will now describe the three main qualities that a specification must have.

Rigor. A specification should not necessarily be an accurate characterization of all the possible reactions of the object. In general, it instead concerns properties verified in all situations. The proper level of detail must be adopted: properties that are too strong excessively constrain the implementation, but too weak constraints only partially depict the semantics of the object. On the other hand, all properties must be perfectly mathematically defined to not give room for subjective interpretation. This is especially true if formal validation techniques are being utilized.

Abstraction. One way to describe the behavior of an object in a rigorous way is to give away its source code. This, however, does not make a good specification. As a matter of fact, code modifications may be necessary in order to improve the efficiency for solving errors, or adapt to a new environment. In all these cases, if the semantics remain the same, there is no reason to change the specification. A specification should be independent of the implementation. For a shared object, the values stored in the local memory of a process and the messages sent, among others, should not be part of the specification.

Simplicity. A specification provides a high level of abstraction to the user of the software building block. What is desirable for the user who makes use of a stack-based data structure is that the value that is popped out be always the last stacked value that has not yet been popped out. Whether or not the implementation uses arrays or linked lists is indifferent. A human user must therefore be able to understand as quickly as possible what the function of the software building block is by only reading its specification, hence the importance of the specification being designed from the developer's perspective who makes use of the object.

For designers, the specification is a goal to be attained, rather than just a way to describe what they have managed to achieve (in the other case, this is instead referred to as *modeling*). The specification is therefore a top-down process: users declare what they require and designers implement what is requested from them. A high-level programming language is an example of a specification method, in which users are the programmers and the designers' role is achieved by the compiler. In the best case, a good specification system can similarly result in a programming language: we will see in Chapter 6 how the concepts presented in this chapter can be integrated within a modern programming language.

Problem. *How can shared objects be specified?*

Specifying a shared object is equivalent to defining which executions are possible and which are not. A specification is thus a set of *concurrent histories* that model the operations allowed by the object. However, a direct description of a set of histories is hardly a reflection of the intuition that one makes of an object. In [SHA 11c], Shavit clearly summarized how shared objects should be specified according to his point of view: "It is infinitely easier and more intuitive for us humans to specify how abstract data structures behave in a sequential setting, where there are no interleavings. Thus, the standard approach to arguing the safety properties of a concurrent data structure is to specify the structure's properties sequentially, and find a way to map its concurrent executions to these 'correct' sequential ones."

Approach. *In this chapter, we define a framework whereby the specification of shared objects is separated into two distinct and complementary facets: a* sequential specification *which describes the functional aspects of the object and a* consistency criterion *which describes how concurrency affects its behavior.*

The sequential specification describes functional aspects of the object without taking into account the constraints related to concurrency. In the case of instant messaging services (whether it be Hangouts, WhatsApp or Skype), the sequential specification simply describes the fact that messages are displayed in the order they are sent. Here, we describe this specification by means of a formal language recognized by a transition system, the *abstract data type*. The *consistency criterion* is a function that transforms sequential specifications into sets of concurrent histories admitted by the shared object. A concurrent history is the modeling of the execution of a program on a given *distributed system*, which describes the computational entities and their calls to the operations of the object that we aim to specify. It assumes the form of a set of events partially ordered by a *process ordering* that describes the sequentiality of the computational entities of the system. In this chapter, apart from in section 1.3.2, we assume that the modeling of the distributed system employing the concurrent stories has already been achieved.

This chapter is organized into three main parts. Section 1.2 presents the concepts of abstract data types and sequential specifications. Then, section 1.3 addresses concurrent histories and more precisely defines wait-free distributed systems. Finally, the focus of section 1.4 is on the notion of consistency criterion.

1.2. Sequential specifications

The sequential specification of an object describes its behavior when it is executed within a sequential environment (in other words, when it is used only by a single process). The works that focus on the specification of data types are tantamount to those describing the properties of formal languages and automata. We also use transition systems to specify the abstract data types. The sequential specification of an abstract data type is the language that it recognizes.

We illustrate this part by focusing on four abstract data types: the integer set (Figure 1.1), the sliding window register (Figure 1.2), the queue (Figure 1.3) and memory (Figure 1.4).

1.2.1. *Abstract data types*

The model that we utilize is a form of transducer which is very similar to a Mealy machine [MEA 55], except that in this case an infinite – but countable – number of states is accepted. The values that may be taken by the data type are encoded in the *abstract state*, taken from a set Z. It is possible to access the object by making use of the symbols of an *input alphabet* A. Contrary to class methods, the input symbols of the abstract data type do not have arguments. As a matter of fact, as a potentially infinite set of operations is allowed, the call to the same method with different arguments is encoded with different symbols. For example, the set (Figure 1.1) has as many insertion input symbols $I(v)$ as there are elements v to be inserted. These symbols correspond to the same method I called with an argument in an implementation. A method can have two types of effects. On the one hand, it may cause a side effect, which changes the abstract state. The corresponding transition in the transitions system is formalized by a *transition function* τ. On the other hand, methods can have a return value taken from an *output alphabet* B, depending on the state in which they are called, and on an *output function* δ. For example, the *pop* method in a stack removes the element at the top of the stack (this is its side-effect) and returns that element (this is its output). The formal definition of abstract data types is given in definition 1.2.

DEFINITION 1.2.– An *abstract data type* (ADT) is a sextuplet $T = (\mathsf{A}, \mathsf{B}, \mathsf{Z}, \zeta_0, \tau, \delta)$ where:

– A is a countable set called *input alphabet*;

– B is a countable set called *output alphabet*;

– Z is a countable set of *abstract states* and $\zeta_0 \in \mathsf{Z}$ is the *initial state*;

– $\tau : \mathsf{Z} \times \mathsf{A} \to \mathsf{Z}$ is the *transition function*;

– $\delta : \mathsf{Z} \times \mathsf{A} \to \mathsf{B}$ is the *output function*.

We consider abstract data types up to isomorphism because the names of the states are never directly used. In addition, we assume that functions τ and δ are computable by a Turing machine to ensure that the difficulties in the implementation are only related to the management of concurrency. The set of ADTs is denoted by $\mathcal{T}$.

The transition function is complete, which expresses the fact that the object reacts to all the external stresses imposed by the user. It is possible to model the fact that a call is incongruous at a given time (for example, during an attempt to place a white pawn when it is the black player's turn in a game of Go) by a loop on certain states or by an error state. In the ADT set, deleting an object that has never been inserted has no effect. This is supported by the fact that $\tau(\zeta, D(v)) = \zeta$ if $v \notin \zeta$. The transition system is also deterministic, which prevents the modeling of objects such as certain types of heaps that have a method to return any of the objects inserted. It would be possible to extend the model to probabilistic or deterministic ADTs, but we have decided not to do so here, as to not add complexity which is not directly related to the topic under study.

We will now see how the instant messaging services of the experiments in the introduction can be modeled in the form of abstract data types. They include a write operation for sending any message and a read operation that displays the result on the screen. When we are just modeling the contents of the screen, a read only returns the last messages, displayed from the oldest to the most recent. We propose to model the instant messaging services by means of *sliding window registers*, formally defined in Figure 1.2. The size k of the sliding window register represents the number of messages that can be displayed on the screen. Sending a message, like writing to the sliding window register, places the argument at the end of the message queue and the read returns the k last written values.

SET

The set (Definition 1.1) is a significant data structure because it is at the basis of most data management systems. It contains elements taken from a countable set of elements Val called support. The abstract data type S_{Val} *offers three types of operations: for each element v of Val, an insert operation I(v) and a delete operation D(v), as well as a comprehensive read operation R that returns all the elements that belong to the set. Initially, the set is empty and when an item is added, it becomes present until it is removed. The figure opposite graphically represents the transition system corresponding to* $S_{\{0,1,2\}}$. *The return values* $\perp$ *of the insertion and deletion operations are not represented there. A transition system path is represented in blue: the word formed by the sequence of the labels on this path is part of the sequential specification of the set.*

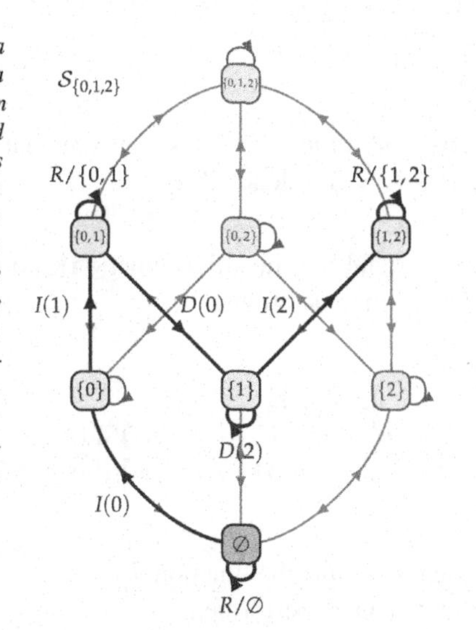

DEFINITION 1.1.– *Let Val be a countable set. The set of support Val is the abstract data type:*

$$\mathcal{S}_{Val} = \left(\mathsf{A} = \bigcup_{v \in Val} \{\mathrm{I}(v); \mathrm{D}(v); \mathrm{R}\}, \mathsf{B} = \mathcal{P}_{<\infty}(Val) \cup \{\perp\}, \mathsf{Z} = \mathcal{P}_{<\infty}(Val), \zeta_0 = \emptyset, \tau, \delta \right)$$

$$\tau : \begin{cases} \mathsf{Z} \times \mathsf{A} & \to & \mathsf{Z} \\ (\zeta, \mathrm{I}(v)) & \mapsto & \zeta \cup \{v\} \\ (\zeta, \mathrm{D}(v)) & \mapsto & \zeta \setminus \{v\} \\ (\zeta, \mathrm{R}) & \mapsto & \zeta \end{cases} \qquad \delta : \begin{cases} \mathsf{Z} \times \mathsf{A} & \to & \mathsf{B} \\ (\zeta, \mathrm{I}(v)) & \mapsto & \perp \\ (\zeta, \mathrm{D}(v)) & \mapsto & \perp \\ (\zeta, \mathrm{R}) & \mapsto & \zeta \end{cases}$$

where $P_{<\infty}(E)$ *designates the set of the finite parts of E.*

Figure 1.1. *Sequential specification of a set*

1.2.2. *Sequential specifications*

The succession of method calls of an object by a sequential program form a sequence of operations admitted by the object. An abstract data type defines the sequential specification of an object, i.e. the set of sequential histories admitted by the object that it describes. The naive way to describe the language recognized by an ADT is to consider the set of infinite paths that traverse its transitions system. This set is too restrictive in the context of this work. In the *sequential specification* (definition 1.5), we allow the knowledge of histories to be partial. Therefore, we accept finite histories, which are the prefixes of infinite histories, as well as histories in which some output values are unknown. The sequential specification is therefore a language of finite and infinite words on an *operation* alphabet (definition 1.4) comprising the input symbols sometimes associated with an output symbol.

DEFINITION 1.4.– Let $T = (\mathsf{A}, \mathsf{B}, \mathsf{Z}, \zeta_0, \tau, \delta)$ an abstract data type.

An *operation* of T is an element of $\Sigma = \mathsf{A} \cup (\mathsf{A} \times \mathsf{B})$. A pair $(\alpha, \beta) \in \mathsf{A} \times \mathsf{B}$ is denoted α/β.

We extend the transition function τ to the operations by applying τ to the input symbol of the operations:

$$\tau_T : \begin{cases} \mathsf{Z} \times \Sigma & \rightarrow \quad \mathsf{Z} \\ (\zeta, \alpha) & \mapsto \tau(\zeta, \alpha) \text{ if } \alpha \in \mathsf{A} \\ (\zeta, \alpha/\beta) & \mapsto \tau(\zeta, \alpha) \text{ if } \alpha/\beta \in \mathsf{A} \times \mathsf{B} \end{cases}$$

We also define the function δ_T^{-1} that returns the set of states of T, in which a given operation can be performed:

$$\delta_T^{-1} : \begin{cases} \Sigma & \rightarrow & \mathcal{P}(\mathsf{Z}) & \\ \alpha & \mapsto & \mathsf{Z} & \text{if } \alpha \in \mathsf{A} \\ \alpha/\beta & \mapsto & \{\zeta \in \mathsf{Z} : \delta(\zeta, \alpha) = \beta\} & \text{if } \alpha/\beta \in \mathsf{A} \times \mathsf{B} \end{cases}$$

COMMENT 1.1.– A symbol of A can be referred to, depending on the context, by the terms "operation", "method" and "input alphabet symbol". When there is no ambiguity, we will favor the term "operation" (in particular to describe algorithms), relatively less connoted to programming languages than the term "method" and more explicit than the term "input alphabet symbol".

DEFINITION 1.5.– A finite or infinite sequence $\sigma = (\sigma_i)_{i \in D} \in \Sigma^\infty$, where D is either $\mathbb{N}$ or $\{0, ..., |\sigma| - 1\}$, is a *sequential history* of an ADT T if there is a sequence of the same length $(\zeta_{i+1})_{i \in D} \in \mathsf{Z}^\infty$ of states of T (ζ_0 being already defined as the initial state) such that for any $i \in D$,

– the eventual output symbol of σ_i is compatible with ζ_i: $\zeta_i \in \delta_T^{-1}(\sigma_i)$;

– the execution of σ_i causes the state ζ_i to change into the state ζ_{i+1}: $\tau_T(\zeta_i, \sigma_i) = \zeta_{i+1}$.

The *sequential specification* of T, denoted by $L(T)$, is the set of its sequential histories.

SLIDING WINDOW REGISTER

Sliding window registers are an important part of this book because they are very often used as examples. A sliding window register of size k, W_k (Definition 1.3), is capable of a write operation $w(n)$ for all $n \in N$ and a read operation r that returns the sequence of the last k values written if they exist, the missing values being replaced with the default 0.
The figure opposite shows part of the transition system of a sliding window register of size 2. For clarity reasons, the transitions leaving and entering a state have been grouped together. After writing the 1, the state (0, 1) consists of the single value 1 preceded by the default value 0. After writing a 2, the first value is shifted to make room for the 2 and the sliding window register shifts to the state (1, 2). If we write a 0, the oldest written value, 1, then leaves the window and the state becomes (2, 0).

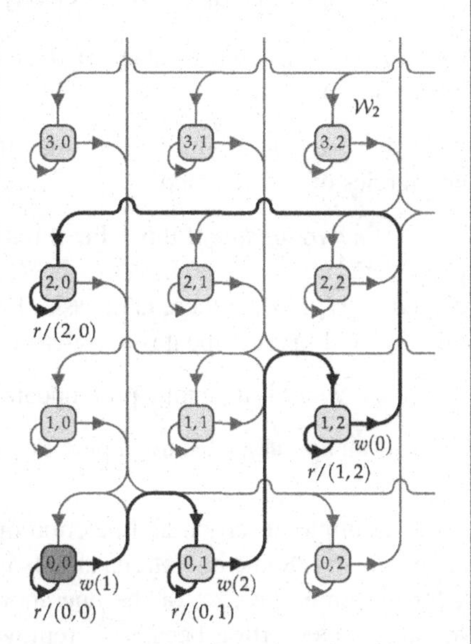

DEFINITION 1.3.– *Let $k \in N$. The integer sliding window register of size k is the ADT :*

$$\mathcal{W}_k = \left(\mathsf{A} = \{r, w(n) : n \in \mathbb{N}\}, \mathsf{B} = \mathbb{N}^k \cup \{\bot\}, \mathsf{Z} = \mathbb{N}^k, \zeta_0 = (0, ..., 0), \tau, \delta\right)$$

$$\tau : \begin{cases} \mathsf{Z} \times \mathsf{A} & \to & \mathsf{Z} \\ ((n_1, ..., n_k), w(x)) & \mapsto & (n_2, ..., n_k, x) \\ ((n_1, ..., n_k), r) & \mapsto & (n_1, ..., n_k) \end{cases}$$

$$\delta : \begin{cases} \mathsf{Z} \times \mathsf{A} & \to & \mathsf{B} \\ ((n_1, ..., n_k), w(x)) & \mapsto & \bot \\ ((n_1, ..., n_k), r) & \mapsto & (n_1, ..., n_k) \end{cases}$$

Figure 1.2. *Sequential specification of the sliding window register*

1.2.3. *Updates and queries*

It is often useful to classify operations according to their effects. The side effect of an operation will be called its *update* part and the link between its state and its return value will be called its *query* part. Certain operations which only have one query or update part are referred to as *pure*. Definition 1.8 thus classifies operations into four categories: update operations that modify the abstract state, query operations whose return value depends on the abstract state, as well as pure update (resp. pure query) operations that are not query (respectively update) operations.

DEFINITION 1.8.– Let an abstract data type $T = (\mathsf{A}, \mathsf{B}, \mathsf{Z}, \zeta_0, \tau, \delta)$ and an input symbol $\alpha \in \mathsf{A}$.

– α is a *pure query* if it has no side effects, i.e. if $\forall \zeta \in \mathsf{Z}, \tau(\zeta, \alpha) = \zeta$. The set of pure queries of T is denoted by $\hat{Q}_T$.

– α is a *pure update* if the value it returns, called its *dummy* value, is independent of the state, i.e. if $\exists \beta \in \mathsf{B}, \forall \zeta \in \mathsf{Z}, \delta(\zeta, \alpha) = \beta$. The set of pure updates of T is denoted by $\hat{U}_T$. When there is no risk of ambiguity between α/β and α, we omit the return symbol in pure updates in order to simplify the notation.

– α is a *query* if it is not a pure update, if $\alpha \in Q_T = \mathsf{A} \setminus \hat{U}_T$.

– α is an *update* if it is not a pure query, if $\alpha \in U_T = \mathsf{A} \setminus \hat{Q}_T$.

For example, insertion and deletion operations of the set are pure updates and R is a pure query. Some operations are also neither pure queries nor pure updates. This is, for example, the case of the operation *pop* of the queue (Figure 1.3) which has both an update portion because it removes the head of the queue and a query part because it returns that element. Operations that are both pure queries and pure updates are generally introspection operations that return information about the type itself, such as the `getClass` method in Java. These operations are less interesting to study because they do not raise any concurrency problems.

An *Update-Query Abstract Data Type* (UQ-ADT) T is a particular ADT for which each operation is either a pure query or a pure update, i.e. $\hat{U}_T \cup \hat{Q}_T = \mathsf{A}$ and $U_T \cap Q_T = \emptyset$. The set is an example thereof. The set of UQ-ADT is denoted by $\mathcal{T}_{UQ}$.

It is always possible to transform any ADT into a UQ-ADT by separating the query part from the update part in operations that both occur at the same time. Revisiting the example of the queue, the operation *pop* can be split into an operation *hd* that solely returns the head of the queue without altering it (pure query) and an operation *rh* that removes the head of the queue without returning its value (pure update). Definition 1.9 explains a generic transformation procedure. The operations α that correspond both to a query and an update are split therein into a pure query α^Q and a pure update α^U.

QUEUE

A queue *is a data structure designed to contain elements of a set of values X, sorted according to a "first in, first out" strategy (*FIFO*). It is specified by an abstract data type (Definition 1.6), having two kinds of operations. An element $x \in X$ can be inserted at the end of the queue using the write operation push(x). The operation pop allows the oldest element that is still present to be retrieved and removed. Initially, the queue is empty. We choose the convention that, in this state in which it is not possible to read the head, the method pop does not alter the state and returns the same dummy value as push(x).*

The queue is not a UQ-ADT because the operation pop modifies the state and returns a value. Definition 1.7 expresses the corresponding UQ-ADT, Q'_X, which splits pop into pure query operation hd (for "head") which returns the head of the queue and a pure update operation rh(x) (for "remove head") which removes the head of the queue if and only if it is equal to x.

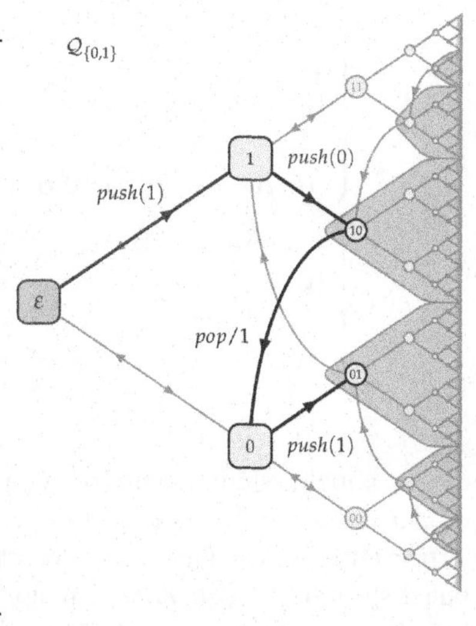

DEFINITION 1.6.– *Let X be a countable set of elements and $\bot$ an element not belonging to X. We denote ε the word of length 0. The type Q_X is the ADT :*

$$\mathcal{Q}_X = (\mathsf{A} = \{push(x) : x \in X\} \cup \{pop\}, \mathsf{B} = \{\bot\} \cup X, \mathsf{Z} = X^\star, \zeta_0 = \varepsilon, \tau, \delta)$$

$$\tau : \begin{cases} \mathsf{Z} \times \mathsf{A} & \rightarrow & \mathsf{Z} \\ (\zeta, push(x)) & \mapsto & \zeta \cdot x \\ (x \cdot \zeta, pop) & \mapsto & \zeta \\ (\varepsilon, pop) & \mapsto & \varepsilon \end{cases} \qquad \delta : \begin{cases} \mathsf{Z} \times \mathsf{A} & \rightarrow & \mathsf{B} \\ (\zeta, push(x)) & \mapsto & \bot \\ (x \cdot \zeta, pop) & \mapsto & x \\ (\varepsilon, pop) & \mapsto & \bot \end{cases}$$

DEFINITION 1.7.– *Let X be a countable set of elements and $\bot$ an element not belonging to X. The type Q'_X is the ADT :*

$$\mathcal{Q}'_X = (\mathsf{A} = \{push(x), rh(x) : x \in X\} \cup \{hd\}, \mathsf{B} = \{\bot\} \cup X, \mathsf{Z} = X^\star, \zeta_0 = \varepsilon, \tau, \delta)$$

$$\tau : \begin{cases} \mathsf{Z} \times \mathsf{A} & \rightarrow & \mathsf{Z} & \\ (\zeta, push(x)) & \mapsto & \zeta \cdot x & \\ (x \cdot \zeta, rh(x)) & \mapsto & \zeta & \\ (y \cdot \zeta, rh(x)) & \mapsto & y \cdot \zeta & \text{if } x \neq y \\ (\varepsilon, rh(x)) & \mapsto & \varepsilon & \\ (\zeta, hd) & \mapsto & \zeta & \end{cases} \qquad \delta : \begin{cases} \mathsf{Z} \times \mathsf{A} & \rightarrow & \mathsf{B} \\ (\zeta, push(x)) & \mapsto & \bot \\ (\zeta, rh(x)) & \mapsto & \bot \\ (x \cdot \zeta, hd) & \mapsto & x \\ (\varepsilon, hd) & \mapsto & \bot \end{cases}$$

Figure 1.3. *Sequential specification of the queue*

DEFINITION 1.9.– Let an ADT $T = (\mathsf{A}, \mathsf{B}, \mathsf{Z}, \zeta_0, \tau, \delta) \in \mathcal{T}$. The corresponding UQ-ADT is $T_{UQ} = (\mathsf{A}', \mathsf{B}', \mathsf{Z}, \zeta_0, \tau', \delta')$, with:

$$\mathsf{A}' = \hat{U}_T \cup \hat{Q}_T \cup \{\alpha^U, \alpha^Q : \alpha \in U_T \cap Q_T\} \mathsf{B}' = \mathsf{B} \cup \{\perp\}$$

$$\tau' : \begin{cases} \mathsf{Z} \times \mathsf{A}' \to & \mathsf{Z} \\ (\zeta, \alpha^U) \mapsto \tau(\zeta, \alpha) \\ (\zeta, \alpha^Q) \mapsto & \zeta \\ (\zeta, \alpha) \ \ \mapsto \tau(\zeta, \alpha) \text{ if } \alpha \in \hat{U}_T \cup \hat{Q}_T \end{cases}$$

$$\delta' : \begin{cases} \mathsf{Z} \times \mathsf{A}' \to & \mathsf{B}' \\ (\zeta, \alpha^U) \mapsto & \perp \\ (\zeta, \alpha^Q) \mapsto \delta(\zeta, \alpha) \\ (\zeta, \alpha) \ \ \mapsto \delta(\zeta, \alpha) \text{ if } \alpha \in \hat{U}_T \cup \hat{Q}_T \end{cases}$$

In a sequential system, the use of an operation that combines updates and queries is equivalent to using the pure query operation followed by the corresponding pure update operation, since these two operations are applied to the same state. In a distributed system, the equivalence is not assured: a concurrent operation could be interleaved between the query and the update. Weak consistency criteria cannot guarantee the atomicity of the operation. Separating the query portion from the update portion can therefore make it possible to give the control of concurrency back to the user of the object. We will return to this discussion with the example of the queue in section 4.4.

1.2.4. *ADT composition*

It is uncommon to only use a single object in a program. In a sequential system, histories of different objects interleave to form a new history that contains the operations of all objects. In a distributed system, the joint use of several objects is harder to specify. Instead of further complexifying the model by considering the utilization of a set of ADTs, we model the set of objects as a single instance of a complex ADT composed of the abstract types of each of the objects. For example, a database management system will not give access to several independent tables, but to a single database composed of all the tables. Similarly, the memory abstract data type (Figure 1.4) is a composition of registers. The composition of two ADTs T_1 and T_2 is the parallel and asynchronous product of their transition systems, formally defined by definition 1.12.

REGISTER AND MEMORY

Shared memory is certainly the most studied object in the literature, because an important branch of the theory of distributed algorithms concerns shared memory models. A memory M_X is a collection of independent registers, each with a different name in X (Definition 1.11).
For each $x \in X$, M_x is the integer register named x. It has an infinite number of write operations of the form $w_x(n)$, where $n \in N$ is the value that we want to write. Its single read operation, r_x, returns the last value written, if it exists, or the default value 0 otherwise. Here, we only consider the register that contains integer numbers and initialized to 0, but it would not be too complicated to extend Definition 1.10 to others data types. The registry is therefore similar to the sliding window register of size 1, only the name of the operations is slightly modified.

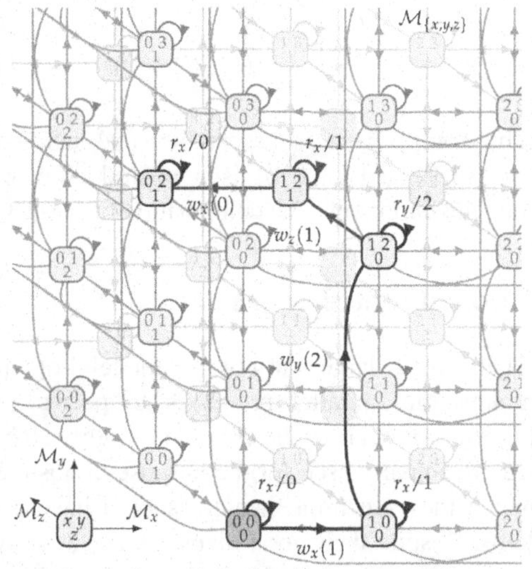

DEFINITION 1.10.– *Let x be a* register *name. The* integer register *on x is the UQ-ADT:*

$$\mathcal{M}_x = (\mathsf{A} = \{\mathrm{w}_x(n) : n \in \mathbb{N}\} \cup \{\mathrm{r}_x\}, \mathsf{B} = \mathbb{N} \cup \{\bot\}, \mathsf{Z} = \mathbb{N}, \zeta_0 = 0, \tau, \delta)$$

$$\tau : \begin{cases} \mathsf{Z} \times \mathsf{A} & \to & \mathsf{Z} \\ (\zeta, \mathrm{w}_x(n)) & \mapsto & n \\ (\zeta, \mathrm{r}_x) & \mapsto & \zeta \end{cases} \qquad \delta : \begin{cases} \mathsf{Z} \times \mathsf{A} & \to & \mathsf{B} \\ (\zeta, \mathrm{w}_x(n)) & \mapsto & \bot \\ (\zeta, \mathrm{r}_x) & \mapsto & \zeta \end{cases}$$

DEFINITION 1.11.– *Let X be a countable set of register names. The integer memory of X is the composition of registers of X, i.e. the UQ-ADT:*

$$\mathcal{M}_X = \prod_{x \in X} \mathcal{M}_x$$

Figure 1.4. *Sequential specification of the register and the memory*

DEFINITION 1.12.– The composition of two ADTs $T = (\mathsf{A}, \mathsf{B}, \mathsf{Z}, \zeta_0, \tau, \delta)$ and $T' = (\mathsf{A}', \mathsf{B}', \mathsf{Z}', \zeta_0', \tau', \delta')$ is the ADT

$$T \times T' = (\mathsf{A} \sqcup \mathsf{A}'^1, \mathsf{B} \cup \mathsf{B}', \mathsf{Z} \times \mathsf{Z}', (\zeta_0, \zeta_0'), \tau'', \delta'')$$

1 $\sqcup$ designates the disjoint union (see notation table).

$$\tau'' : \begin{cases} (\mathsf{Z} \times \mathsf{Z}') \times (\mathsf{A} \sqcup \mathsf{A}') \to & \mathsf{Z} \times \mathsf{Z}' \\ ((\zeta, \zeta'), \alpha) & \mapsto (\tau(\zeta, \alpha), \zeta') \text{ if } \alpha \in \mathsf{A} \\ ((\zeta, \zeta'), \alpha') & \mapsto (\zeta, \tau'(\zeta', \alpha')) \text{ if } \alpha' \in \mathsf{A}' \end{cases}$$

$$\delta'' : \begin{cases} (\mathsf{Z} \times \mathsf{Z}') \times (\mathsf{A} \sqcup \mathsf{A}') \to & \mathsf{B} \cup \mathsf{B}' \\ ((\zeta, \zeta'), \alpha) & \mapsto \delta(\zeta, \alpha) \text{ if } \alpha \in \mathsf{A} \\ ((\zeta, \zeta'), \alpha') & \mapsto \delta'(\zeta', \alpha') \text{ if } \alpha' \in \mathsf{A}'. \end{cases}$$

The composition is associative and commutative (up to isomorphism), and $T_1 = (\emptyset, \emptyset, \{\zeta_0\}, \zeta_0, \emptyset, \emptyset)$ is a neutral element. The structure $(\mathcal{T}, \times)$ is thus a commutative monoid. The composition of two UQ-ADT is a UQ-ADT.

1.3. Concurrent histories

There is a large variety of models of distributed systems, ranging from communicating parallel processes to peer-to-peer networks, including dynamic architectures that allow the creation of lightweight threads on-the-fly. Shared objects, which are actually used for the purpose of abstracting the complexity of the systems in which they are implemented, should have a specification independent of these systems. Despite their great diversity, all these systems have one thing in common: the events that occur within them are not totally ordered. In general, computational entities (process, peers, execution threads, etc.) are sequential, which imposes an ordering between their own events, yet concurrent events are not ordered. For example, in the case of communicating sequential processes: an event a precedes an event b if they are executed by the same process in that order. We first present our model of concurrent histories, we then apply it to systems composed of sequential processes communicating by messages.

1.3.1. *Definition*

A concurrent history is an abstract model of the actual executions that may occur. The manner in which the executions are modeled is left to the system designer and depends on the system under consideration. In addition, modeling choices (for example, should modeling take real time into account?) may lead to slightly different properties. A *concurrent history* (definition 1.14) is simply a set of *events*, partially ordered by a *process order* and labeled by *operations*. In our model, operations are typically those of an abstract data type (but not necessarily exactly) of the form α/β or α. In addition, the process order is imposed to verify the two following properties (this is referred to as *well partial order*, see definition 1.13):

Well-founded relation. There is no infinite sequence of elements strictly decreasing according to the process order. In our model, this means that every execution has a temporal origin: there is no history that has always existed or process infinitely fast that can produce an infinite number of events in a finite time.

Lack of infinite antichain. In any infinite set of events, there is at least a couple of ordered elements according to the process order. In our model, this means that at any time, the number of entities in the system is finite, although not necessarily bounded.

DEFINITION 1.13.– Let E be a set and $\leq$ a partial order on E. The relation $\leq$ is a *well partial order* if, for every sequence $(e_n)_{n\in\mathbb{N}}$ of elements of E, there exist i and j such that $i < j$ and $e_i \leq e_j$.

DEFINITION 1.14.– A *concurrent history* is a quadruplet $H = (\Sigma, E, \Lambda, \mapsto)$ whose various fields are detailed below.

– Σ is a countable set of *operations*;

– E is a countable set of *events*. For a history H, the accessor E_H allows access to the event set of H;

– $\Lambda : E \to \Sigma$ is the *labeling function* of the events by operations;

– $\mapsto$, the *process order* , is a well partial order on E.

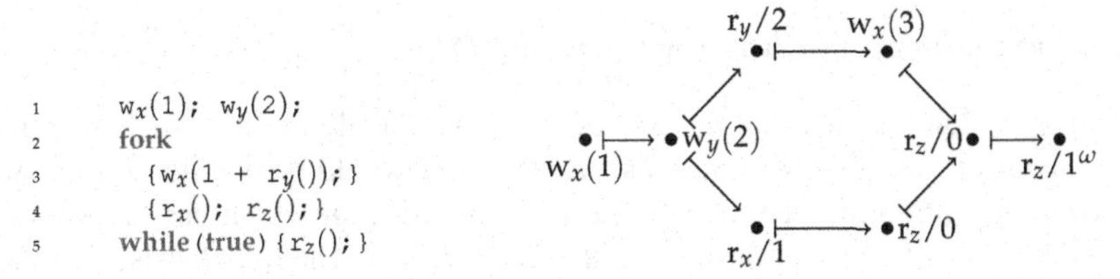

(a) Concurrent program. (b) History corresponding to an execution.

Figure 1.5. *Graphic representation of concurrent histories. The concurrent history of Figure 1.5(b) is an example of the execution of the program of Figure 1.5(a). The points • represent the events that have occurred during execution, labeled by updates and queries operations on the shared registers x, y and z. The dummy return value of pure updates is not represented so as to not add further complexity. The events are connected by arrows $\mapsto$ that describe the direct process order, whose process order is the transitive closure. At each call to fork, the process is split into two threads of execution that join further on. The history is therefore not linear. The notation ω as exponent designates an infinite chain of events (depending on $\mapsto$) labeled by the same operation*

In a system made up of a set of communicating processes, two events are ordered by $\mapsto$ if, and only if, they are generated by the same process. Each process thus corresponds to a maximal chain of events of H, i.e. a maximal subset of E totally ordered by $\mapsto$. We will denote $\mathscr{P}_H$ as the set of the maximal chains of H and we will make use of the term "process" to designate such a chain, even in the models that are not based on a collection of processes.

In addition, we extend the notations of (pure) queries and (pure) updates of T to the events of H: $\hat{U}_{T,H} = \{e \in E_H : \Lambda(e) \in \hat{U}_T\}$, and similarly for $\hat{Q}_{T,H}$, $U_{T,H}$ and $Q_{T,H}$.

Figure 1.5 introduces the manner in which we graphically represent concurrent histories.

Complex distributed systems can be modeled by concurrent histories. A concurrent history can be perceived as a structure that concisely encodes several sequential histories, each being a linearization of the concurrent history (definition 1.15). A linearization is thus a sequence of operations labeling events in the history, in an order that respects the process order.

DEFINITION 1.15.– Let $H = (\Sigma, E, \Lambda, \mapsto)$ be a concurrent history. A *linearization* of H is a word (finite or infinite) $l = \Lambda(e_0)\ldots\Lambda(e_n)\ldots$ that contains all of the events of E, i.e. $\{e_0, \ldots, e_n, \ldots\} = E$, in an order that respects the process order, i.e. $e_i \mapsto e_j \Rightarrow i < j$.

The linearization set of H is denoted by $\text{lin}(H)$.

COMMENT 1.2.– The fact that the process order is a well partial order is important for linearizations to exist. Without this hypothesis, it would be possible for an event to have an infinite past according to the process order, in which case it would be impossible to give it a finite position in the linearization. The hypothesis of the well partial order will be necessary again in Chapter 5 to prove Theorem 5.10.

We now present our projection operator, the all-purpose tool that will allow us to manipulate concurrent histories to define consistency criteria. The projection is formally defined in definition 1.16. Figure 1.6 illustrates the operation of this projection. Let $\dashrightarrow$ be a well partial order and A, B the two subsets of events. The projection $H^{\dashrightarrow}[\mathsf{A}/\mathsf{B}]$ preserves the updates of A and the queries of B and reorders them according to $\dashrightarrow$. More specifically, it has the following three actions:

– it deletes the events that are not in A;

– it hides the return value of the operations that are not in B. This means that we only consider the read part of the operations labeling the events of B. In general, we will make sure that $\mathsf{A} \subset \mathsf{B}$. According to definition 1.16, $H^{\dashrightarrow}[\mathsf{A}/\mathsf{B}] = H^{\dashrightarrow}[\mathsf{A} \cup \mathsf{B}/\mathsf{B}]$;

– it reorders the events in the history according to $\dashrightarrow$. This allows the addition or suppression of dependencies between events. The events of $H^{\dashrightarrow}[\mathsf{A}/\mathsf{B}]$ are those of H, ordered according to $\dashrightarrow$. There is therefore a correspondence between the linearizations l of H and the linear extensions $\dashrightarrow$ of the process order, such that $\text{lin}\left(H^{\dashrightarrow}[E_H/E_H]\right) = \{l\}$.

DEFINITION 1.16.– Let $H = (\Sigma, E, \Lambda, \mapsto)$ be a concurrent history, two subsets of events $\mathsf{A}, \mathsf{B} \subset E_H$ and a well partial ordering $\dashrightarrow$ on E_H. The *projection* $H^{\dashrightarrow}[\mathsf{A}/\mathsf{B}]$ is the concurrent history $(\Sigma', E', \Lambda', \mapsto')$ whose various fields are described hereafter:

– $E' = \mathsf{A} \cup \mathsf{B}$;

– if $e \notin \mathsf{B}$ and $\Lambda(e) = \alpha/\beta$, then $\Lambda'(e) = \alpha$. Otherwise, $\Lambda'(e) = \Lambda(e)$;

– $\Sigma' = \Lambda'(E')$;

– $\mapsto'$ is the restriction of $\dashrightarrow$ to E'.

When the three fields are not necessary, we are allowed to use the simplified notations $H^{\dashrightarrow} = H^{\dashrightarrow}[E_H/E_H]$ and $H[\mathsf{A}/\mathsf{B}] = H^{\mapsto}[\mathsf{A}/\mathsf{B}]$.

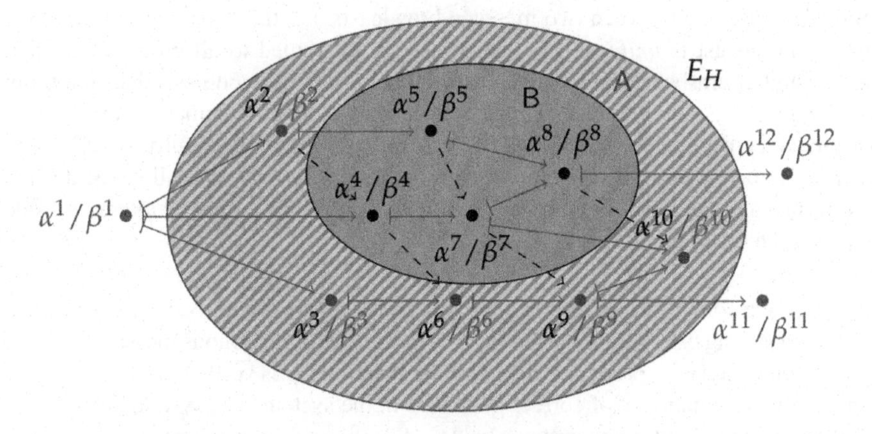

Figure 1.6. *$H^{\dashrightarrow}[\mathsf{A}/\mathsf{B}]$ contains the updates of* A *and the queries of* $\mathsf{A} \cup \mathsf{B}$. *For the color version of this figure, see www.iste.co.uk/perrin/distributed.zip*

1.3.2. *Asynchronous processes communicating by messages*

We now more precisely define asynchronous distributed systems communicating by messages, and in particular wait-free distributed systems.

The computation system is composed of n processes $p_0, ..., p_{n-1}$. Each process executes sequentially and has a local memory that is considered unbounded. The processes are asynchronous, i.e. there is no bound on the runtime of their basic primitives, neither relative to the execution time of the primitives by the other processes, nor to that of the other calls to the primitives by the same process.

Processes communicate by exchanging messages through a complete network of communication channels. To this end, processes have access to message broadcast and reception primitives. A broadcast message is sent to all other processes, including the issuer. The channels are asynchronous, in other words, there is no bound on the time between the broadcast and the reception of a message. For example, a broadcast message can be received very quickly by one process and after a very long time period by another. During the broadcasting of a message, the sending process receives it immediately (we can easily notice that the broadcast primitive locally calls the algorithm for receiving a message).

Among the n processes, up to $t \leq n$ can be *faulty*. A faulty process simply stops operating: at a certain point, called *crash failure*, it can no longer broadcast nor receive messages or execute any action. Crash failures may occur during the execution of an action, for example between two message broadcasts. On the other hand, we assume that the broadcast is *uniform*, i.e. that messages are routed to all processes even if the transmitter crashes during a broadcast (see below). This hypothesis does not bring any computing power to the system: the uniformity can be implemented in a system not guaranteeing this property [HAD 13]. A process that is not faulty is said to be *correct*. A correct process, a faulty process before the failure, normally executes the algorithms, and the local memory utilized by the algorithm cannot be altered outside its execution (even by local action).

Let φ be a condition or a set of conditions on the maximal number t of failures that may occur during execution. We denote $AS_n[\varphi]$ as the asynchronous message-passing system composed of n processes in which the condition φ is verified. In $AS_n[t < \frac{n}{2}]$, there is always a majority of correct processes in the system. The system $AS_n[t \leq 1]$ allows one single fault in the system. In $AS_n[t = 0]$, no fault may occur.

In $AS_n[\emptyset]$, there is no condition on the number of faults. All processes are thus likely to crash during execution. The system $AS_n[\emptyset]$ is called *wait-free asynchronous message-passing distributed system* or simply *wait-free system*.

A process is subject to two kinds of interaction: calls to the operations of an abstract data type and message reception. The behavior of the processes at these moments is defined by an algorithm depending on the implemented object. The execution of this algorithm is indivisible: an algorithm cannot receive a message from another process (remember that its own messages are immediately received) nor perform an operation of the shared object when addressing another message or another operation.

In order to broadcast a message m, a process has three communication primitives: `broadcast`(m), `FIFO broadcast`(m) and `causal broadcast`(m) specified by the following properties:

1) The primitive `broadcast`, usually called *reliable broadcast* [HAD 13], allows a response to the situation in which a transmitter crashes during broadcast and provides the following two properties:

Validity. If a process receives a message m at least once, then m has been broadcast at least as many times by a process.

Uniformity. If a correct process receives a message m, then all correct processes receive m.

These properties imply that any message broadcast by a correct process will be received by all correct processes. It turns out that if a correct process broadcasts a message, it will immediately receive it locally and since it is correct, all other correct processes will also receive it following the uniformity property.

2) The primitive `broadcast` does not guarantee the order of the messages. The primitive `FIFO broadcast` guarantees the two previous properties, as well as an additional property that ensures that two messages broadcast by the same process will be received in the order that they were sent.

FIFO reception. If a process broadcasts a message m and then a message m' using `FIFO broadcast`, then no process will receive m' before m.

3) Finally, `causal broadcast` guarantees the two properties of the primitive `broadcast` as well as that Lamport's "happened-before" relation [LAM 78] be respected. Note that this property implies that of the primitive `FIFO broadcast.`

Causal reception. If a process receives a message m and then broadcasts a message m' using `causal broadcast`, then no process will receive m' before m.

In practice, these three primitives can be implemented each based on the others in $AS_n[\emptyset]$ [BIR 87, RAY 91, RAY 96, SCH 89]. Therefore, they do not contribute to computing power but only with ease of use. On the other hand, they do not have the same costs. Thus, we prefer to use a weaker primitive when the additional assumptions are not necessary.

1.3.3. *Modelization as of concurrent histories*

Due to the failures that may occur during execution, there is no unique method capable of modeling execution by means of concurrent histories. Furthermore, if we consider the case of a crash failure occuring when a process executes an operation, if the operation had just started, it could not have had any effect on other processes. Conversely, if the operation was about to finish, the operation should be taken into account. In the model of an execution, this kind of operation must be taken into account, either completely or not at all. More accurately, a concurrent history $H = (\Sigma, E, \Lambda, \mapsto)$ models the execution of a program utilizing an abstract data type $T = (\mathsf{A}, \mathsf{B}, \mathsf{Z}, \zeta_0, \tau, \delta)$ in a wait-free system if:

– the set of operations is $\Sigma = \mathsf{A} \cup (\mathsf{A} \times \mathsf{B}')$, where B' contains B and all the output values that can be returned by the implementation of T, including error messages;

– E only contains calls to the object operations that occurred during the execution including all those that have been completed;

– for any $e \in E$, $\Lambda(e) = \alpha/\beta$, where α is the identifier of the method called and β is the value returned when executing e;

– for any $e, e' \in E$, $e \mapsto e'$ if, and only if, e and e' are different actions of the same process and e occurred before e'.

1.4. Consistency criteria

A consistency criterion is a property that makes a link between the *sequential specification* and the *concurrent histories* that it allows. Figuratively speaking, a consistency criterion can be seen as a point of view, one way of looking at a concurrent history to make it appear sequential and in agreement with the sequential specification.

1.4.1. *The consistency criteria set*

Since the sequential specification of an object in a sequential system is the set of the sequential histories that it admits, the specification of a shared object is a set of concurrent histories. A consistency criterion characterizes the concurrent histories admissible for a given ADT. It is thus a function that associates a concurrent specification with each ADT (definition 1.17).

DEFINITION 1.17.– A consistency criterion is a function $C : \mathcal{T} \to \mathcal{P}(\mathcal{H})$. The set of consistency criteria is denoted by $\mathcal{C}$.

An algorithm A_T is C-consistent for a criterion $C \in \mathcal{C}$ and an ADT $T \in \mathcal{T}$, if all the operations performed by correct processes terminate[2] and all the executions that it admits can be modeled by a C-consistent history, i.e. which belongs to $C(T)$.

2 The condition that all operations terminate is usually called *wait-free* as well. Other progression conditions could be specified, but we only consider wait-free algorithms here.

Figure 1.7 illustrates our model by formally defining local consistency. Local consistency (LC) is the consistency criterion of local variables that are not shared between processes: every process applies its updates in its local copy and there is no exchange of data between processes.

The consistency criteria space naturally inherits the properties of the power set and, in particular, its lattice structure for inclusion. The ordering $\leq$ expresses the strength of the consistency criterion. If $C_1 \leq C_2$, then C_2 is *stronger* than C_1 because it guarantees stronger properties on the histories it admits. Thus, a C_2-consistent implementation can still be used instead of a C_1-consistent implementation of the same ADT, if $C_1 \leq C_2$. The upper bound, given by the operator $+$, is the *conjunction*. It allows the consistency criteria to be strengthen by jointly making use of them, so as to ensure the properties of both criteria at the same time.

DEFINITION 1.18.– Let $\leq$ be the relation defined on the consistency criteria by $C_2 \leq C_1$, if $\forall T \in \mathcal{T}, C1(T) \subset C2(T)$. The algebraic structure $(\mathcal{C}, \leq)$ is a bounded lattice whose two binary operations are $+$ (upper bound) and $\triangle$ (lower bound), the maximum is $C_\top$ and the minimum $C_\bot$, defined by:

$$C_1 + C_2 : \begin{cases} \mathcal{T} \to & \mathcal{P}(\mathcal{H}) \\ T \mapsto & C_1(T) \cap C_2(T) \end{cases} \qquad C_\top : \begin{cases} \mathcal{T} \to \mathcal{P}(\mathcal{H}) \\ T \mapsto \quad \emptyset \end{cases}$$

$$C_1 \triangle C_2 : \begin{cases} \mathcal{T} \to & \mathcal{P}(\mathcal{H}) \\ T \mapsto & C_1(T) \cup C_2(T) \end{cases} \qquad C_\bot : \begin{cases} \mathcal{T} \to \mathcal{P}(\mathcal{H}) \\ T \mapsto \quad \mathcal{H} \end{cases}$$

1.4.2. *Consistency criteria composition*

Complex programs are in fact often composed of several shared objects. Besides, the main interest of object-oriented programming is the capability to encapsulate software building blocks in classes isolated one from another. A legitimate question is therefore the study of the joint behavior of shared objects, if they are individually consistent. More specifically, let us consider a program composed of instances of two abstract data types T_1 and T_2 that verify the same consistency criterion C.

The expected behavior is that the two objects jointly behave as if there was only one single instance of the composed type $T_1 \times T_2$ with the consistency criterion C. This property is called *locality* (definition 1.20). Locality breaks down into two properties. Composability guarantees that if two objects are consistent, their composition is also consistent. Conversely, decomposability guarantees that if the composition is consistent, then the components are themselves consistent.

LOCAL CONSISTENCY

✓ Composable p.42
✓ Decomposable p.42
✗ Not shared p.142

A local variable contains an object instance that is not shared. The calls to the operations of this instance nonetheless form concurrent histories that have a particular structure, specified by a consistency criterion: local consistency. In local consistency, each process is independent of the others. For example, in the history below, two processes write and read to a set of integers (see Page 29). The first process inserts the number 1 then reads the state $\{1\}$ an infinite number of times. The second process inserts the number 2 then reads the state $\{2\}$ an infinite number of times. All the reads properly contain the values inserted by the process that carries out the read, regardless of what the other process does.

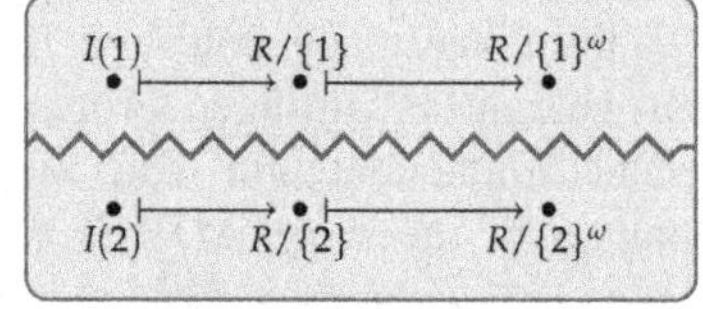

Formally, a history H is locally consistent for an abstract data type T (Definition 1.19) if, for each process $p \in P_H$, the linearization of history $H[p/p]$, containing only the events of p (which is unique because its events are totally ordered), is consistent with the sequential specification $L(T)$ of T. In the above history, the following linearizations of the events of each process are consistent with the sequential specification of the shared object:

$$I(1) \cdot R/\{1\}^{\omega} \qquad\qquad I(2) \cdot R/\{2\}^{\omega}$$

DEFINITION 1.19.– Local consistency *is the consistency criterion:*

$$LC : \begin{cases} \mathcal{T} & \rightarrow \quad \mathcal{P}(\mathcal{H}) \\ T & \mapsto \quad \{H \in \mathcal{H} : \forall p \in \mathscr{P}_H, \mathrm{lin}(H[p/p]) \cap L(T) \neq \emptyset\} \end{cases}$$

The only change in the history opposite compared to the above locally consistent history is the value returned by the reads of the second process starting from the second reading. The second process reads now $\{2\}$ and then $\{1,2\}$ without having inserted 1 between its two reads. The history is not locally consistent because, in the linearization for the second process, the shift from state $\{2\}$ to state $\{1,2\}$ is not consistent with the sequential specification $L(T)$ of T.

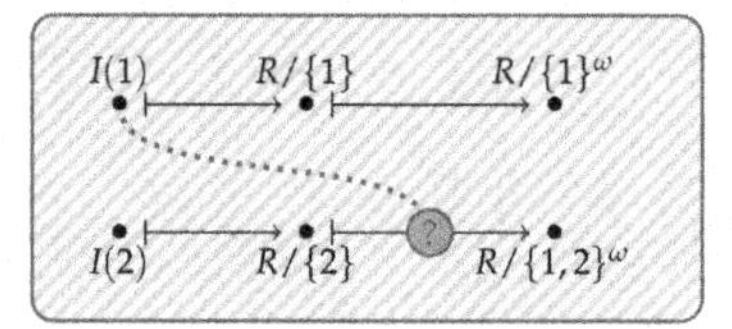

Figure 1.7. *Local consistency*

Local consistency is a local criterion. The similarity in the naming of local consistency and locality does not refer to the same concept: the "locality" of the consistency criterion is spatial (compared with the process in which the operation is called) while that of the property is rather temporal: the term was originally used in [VIT 03] to describe linearizability (see page 29) which draws its composability from its relationship with real time.

DEFINITION 1.20.– Let $C(T_1, T_2)$ be the set of histories whose projections on the events concerning $T_1 \in \mathcal{T}$ and $T_2 \in \mathcal{T}$ are both C-consistent. Formally, a history

H is in $C(T_1, T_2)$ if there is a partition $\{E_1, E_2\}$ of the events of E_H, such that the projections of H upon E_1 and E_2 are C-consistent for each of the two ADTs:

$$C(T_1, T_2) = \left\{ \begin{array}{r} H \in \mathcal{H} : \exists E_1, E_2 \subset E_H,\ E_1 \sqcup E_2 = E_H \\ \wedge\ H[E_1/E_1] \in C(T_1) \\ \wedge\ H[E_2/E_2] \in C(T_2) \end{array} \right\}$$

A criterion $C \in \mathcal{C}$ is *composable* if: $\forall T_1, T_2 \in \mathcal{T}, C(T_1 \times T_2) \supset C(T_1, T_2)$.

A criterion $C \in \mathcal{C}$ is *decomposable* if: $\forall T_1, T_2 \in \mathcal{T}, C(T_1 \times T_2) \subset C(T_1, T_2)$.

A criterion $C \in \mathcal{C}$ is *local* if: $\forall T_1, T_2 \in \mathcal{T}, C(T_1 \times T_2) = C(T_1, T_2)$.

After all, what happens if in the same program we use C-consistent shared objects for abstract data types T_1 and T_2, where C is not composable? The resulting histories will not necessarily be C-consistent for the ADT $T_1 \times T_2$, but they will still verify the following property: their projections on the events of T_1 and T_2 will be C-consistent. The consistency criterion that models these histories is called *composition closure* of C, denoted by $C^\star$. More accurately, definition 1.21 defines the composition closure of C as the upper bound of all composable criteria weaker than C and Proposition 1.1 shows that this is the strongest composable criterion weaker than C. This property has an impact on the difficulty to guarantee locality: if a criterion is not composable, the only means to make it composable by weakening it is to consider the objects independently. This explains why very few of the criteria presented hereafter are composable, and those that are, are very weak.

DEFINITION 1.21.– Let C be a consistency criterion. The *composition closure* of C is the consistency criterion:

$$C^\star = \sum \{C' \leq C : C' \text{ composable}\}$$

PROPOSITION 1.1.– *For all C, $C^\star$ is the strongest composable criterion weaker than C.*

PROOF.– First, $C^\star \leq C$. In effect, let $T \in \mathcal{T}$ and $H \in C(T)$. For any composable criterion $C' \leq C$, $H \in C(T)$, therefore $H \in C'(T)$. Furthermore, $H \in C^\star(T)$, thus $C^\star \leq C$.

Second, $C^\star$ is composable. In effect, let $T_1, T_2 \in \mathcal{T}$, $H_1 \in C^\star(T_1)$, $H_2 \in C^\star(T_2)$ and $H \in H_1 \times H_2$. For any $C' \leq C$ composable, $H_1 \in C'(T_1)$, $H_2 \in C'(T_2)$. On the other hand, C' is composable, thus $H \in C'(T_1 \times T_2)$. Finally, $H \in C^\star(T_1 \times T_2)$; thus, $C^\star$ is composable.

Thirdly, $C^\star$ is stronger than all the other composable criteria weaker than C by construction. □

1.5. Conclusion

In this chapter, we have proposed to split the specification of shared objects into two separate facets. On the one hand, the sequential specification describes the functional aspect of the object operations. On the other hand, the concept of consistency criteria expresses the quality of service when taking concurrency into account. At the beginning of this chapter, we identified the three qualities that a good specification should have. It is now time to confront our proposal.

Rigor. This quality is probably the simplest to obtain: all notions of this chapter have a precise mathematical definition that leaves no room for interpretation. In addition to the basic concepts of abstract data type, concurrent history and consistency criterion, we have defined some basic operators from the competing histories, such as projections and linearizations, which will play an important role in rigorously defining the consistency criteria in the following chapters.

Abstraction. Concurrent histories are themselves high-level system abstractions in which objects can be implemented. Wait-free systems, presented here as an example, will be used hereafter in algorithmic analysis only, but never to define new consistency criteria. Besides, instant messaging services cited as examples typically utilize a server in their implementation. Therefore, the specification of shared objects does not depend on the system in which they are deployed.

Simplicity. Breaking down the specification into two facets allows us to understand each with half the difficulty. This responds to Shavit's observation in [SHA 11c]: "It is infinitely simpler and more intuitive for us humans to specify how abstract data structures behave in a sequential setting." Regarding the sequential specification, we decide on a definition based on transition systems, because this is a tool that is intuitive, central to all branches of computer science, and whose use through modeling and verification tools is widely documented. The complexity is therefore entirely directed towards the consistency criteria, of which each allows an infinite number of different objects to be specified.

Many other models and concepts have been proposed to specify particular shared objects. In the next chapter, we show how our model makes it possible to encompass all these concepts, and to compare their expressions in this common framework.

2

Overview of Existing Models

2.1. Introduction

In Chapter 1, we have proposed that shared objects shall be specified by a sequential specification, which describes the functional aspect of their operations and a consistency criterion that describes how concurrency is taken into account by the object. Obviously, the specification in the presence of concurrency is not a new problem. Four scientific communities, have been addressing this problem and have contributed their own solutions:

Distributed algorithms. The main problems in this area relate to data structures which can be implemented in different systems generally; including shared memory. Most often, the objects targeted in this area hide concurrency by behaving as if all accesses were sequential. It is said that they are verifying *strong consistency*.

Transactional systems. Databases and transactional memories make use of a simple mechanism for implementing complex operations: transactions encapsulate simple operations performed as if they were a single one more complex. Specifying the result of transactions based on the basic operations has been the focus of a lot of research.

Operating systems. Besides the theoretical aspect necessary for the understanding of the computability of distributed systems, distributed algorithms are sometimes criticized for their lack of realism when it comes to the reality of modern systems, such as *cloud*-based architectures, whose scale and failures that may occur render strong consistency unachievable. The targeted applications for these systems, such as collaborative documents editors [NÉD 13, OST 06, PRE 09, WEI 09] and data warehouses [DEC 07, LAK 10] are accessible by several users sometimes geographically remote, and their data are replicated on all the processes or part of them. Strong consistency is usually not verified; instead, these applications guarantee that, if all participants stop modifying the object, all replicas will end up converging into a common state. It is therefore necessary to specify the state to which they converge.

Parallel programming. Similarly, the sheer size of current supercomputers implies that the cost of manufacturing, development and accessing shared memory becomes very significant. Weakly consistent memory models have been proposed [AHA 95, LIP 88] for parallel programming. As algorithms do not operate the same way with different memory models, it is important to accurately specify these new models.

Problem. *What solutions have been proposed by the various communities addressing shared objects to specify them?*

The works of the different communities have different goals, which have driven them to develop different tools and formalisms. As a result, it is very difficult to compare these works without expressing them in a unified framework. To illustrate this point, we shall reuse the example of the behavior observed in the introduction with the WhatsApp instant messaging service (Figure I.2). Eventual consistency is not implemented (messages appear in different orders to both speakers), therefore neither the formalism of distributed algorithms nor that of collaborative systems allows it to be correctly specified; the operations (sending a message and displaying the messages received) do not correspond to reading or writing to memory. Finally, the application was not developed based on transactions, and therefore transactional systems theory does not apply.

Approach. *In this chapter, we aim to identify the key concepts proposed in each community. Whenever possible, we strive to extend the definitions so that they can be inserted in the context of Chapter 1.*

We have been careful with the criteria defined here, so that they coincide with their definitions of the state-of-the-art. For example, pipelined consistency applied to memory in $AS_n[\emptyset]$ allows for the same histories as PRAM memory as in [LIP 88].

We formally define the most important criteria in our model within a form illustrated with the histories given in Figure 2.1. For each of these histories (except the one in Figure 2.1(f)), an instance such as $\mathcal{W}_2 \times \mathcal{S}_{[a-z]}$, which is the composition of a sliding window register of size 2 (definition 1.3) and of a set of letters (definition 1.1) is shared between two processes; the first process writes 1 in the sliding window register (operation $\mathrm{w}(1)$) then inserts a in the set (operation $\mathrm{I}(a)$), while the second process writes 2 in the sliding window register (operation $\mathrm{w}(2)$) then reads the set (operation R). Next, both read the sliding window register an infinite number of times (operation r). In Figure 2.1(f), the insertion in the set is replaced by a read to the sliding window register. The criteria verified by these histories are represented in Figure 2.1(a). The red shaded areas in this figure contain only criteria weaker than a non-composable criterion C, but not its composition closure $C^\star$. According to proposition 1.1, these criteria are therefore not composable.

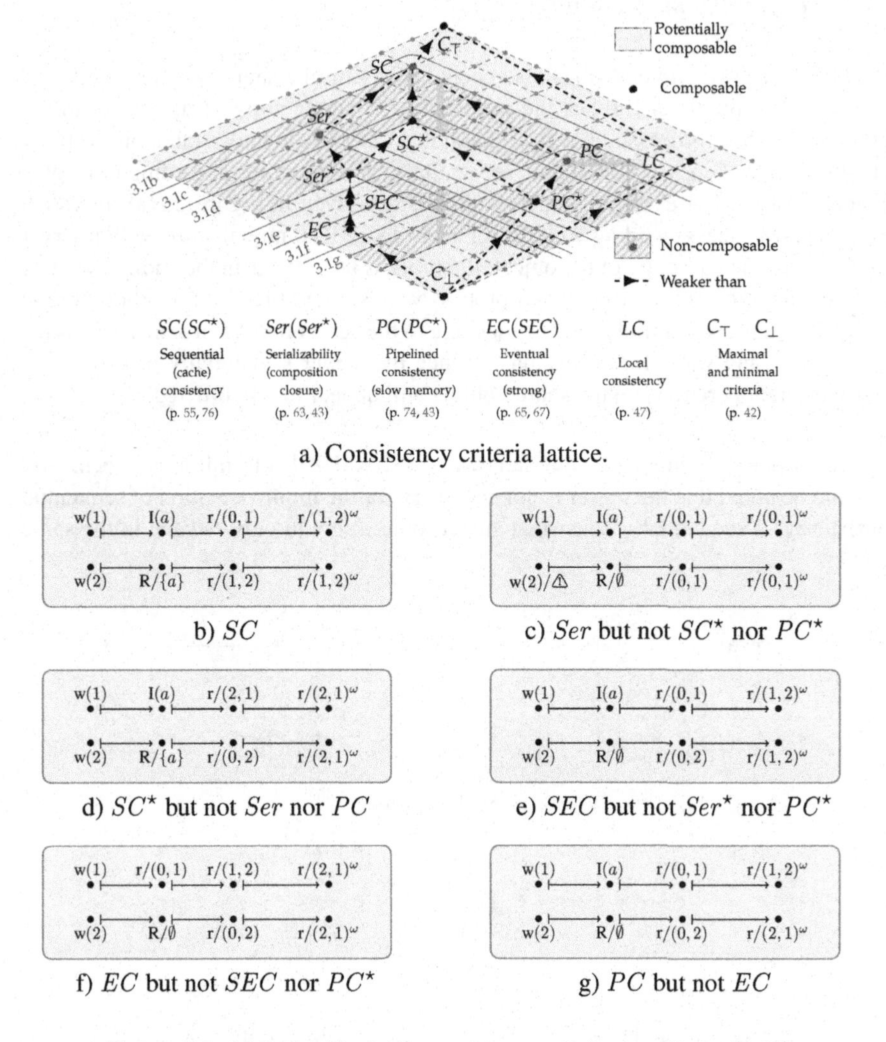

a) Consistency criteria lattice.

b) SC

c) Ser but not $SC^\star$ nor $PC^\star$

d) $SC^\star$ but not Ser nor PC

e) SEC but not $Ser^\star$ nor $PC^\star$

f) EC but not SEC nor $PC^\star$

g) PC but not EC

Figure 2.1. *Histories illustrating various consistency criteria. For the color version of this figure, see www.iste.co.uk/perrin/distributed.zip*

This chapter dedicates a section to each of the main communities that have developed tools for the specification of shared objects. Section 2.2 focuses on strong consistency mainly studied in the theory of distributed algorithms, section 2.3 is dedicated to transactional systems, section 2.4 studies eventual consistency and the implemented means that can guarantee it and section 2.5 presents memory models used in parallel programming.

2.2. Strong consistency

2.2.1. *Sequential consistency*

The simplest way to describe the expected behavior of a shared object is certainly to require that the latter mimics the existence of a single copy of the object in the network, which all processes access sequentially. One way to understand this property is that this shared copy is stored on a server that sorts operations according to its own sequential order. For example, consider the history given in Figure 2.2(a), in which two processes access a sliding window register of size 2 communicating with a server that manages the sole copy of the object; the process p_0 writes 1 in the sliding window register (operation w(1)), the second process writes 2 and then both read (operation r). To write, a process simply sends a message to the server to read. It sends a message to the server and waits for its response. Since the server sequentially processes the messages, the operations on the shared object will appear totally ordered.

The purpose of sequential consistency (SC) [LAM 79] is to mimic this behavior. It should be noted that the server is not necessary for the implementation of sequential consistency; a sequentially consistent shared object should only *behave* in the same manner.

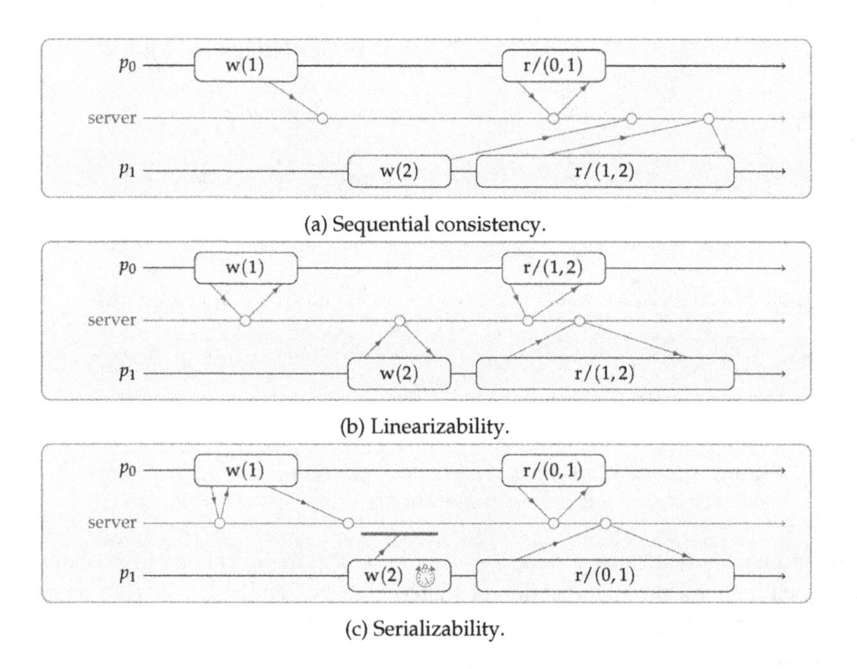

(a) Sequential consistency.

(b) Linearizability.

(c) Serializability.

Figure 2.2. *Strong consistency and serializability with the central server*

More specifically, a concurrent history is sequentially consistent, if there exists a sequential history such that:

– the sequential history contains the same events as the concurrent history;

– two events of the same process appear in the same order in the sequential history;

– the sequential history is correct with regard to the object being considered.

SEQUENTIAL CONSISTENCY AND INFINITE HISTORIES

I(2) I(1) I(3) D(2) I(4) D(3) I(5) D(4) $(\mathrm{I}(n)\cdot \mathrm{D}(n-2))^{n\geq 6}$

R/{1} R/{0}

R/∅ R/∅ R/∅ R/∅ R/∅ R/∅ R/∅ R/∅ $(\mathrm{R}/\emptyset)^{\omega}$

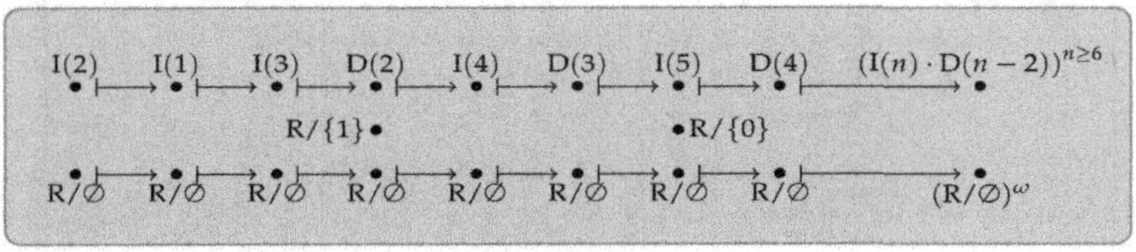

We now focus on the above history and especially on the two processes that execute an infinite number of operations. The first process inserts the values 2, 1 and 3 in a sequentially consistent shared set then alternates between deleting the smallest element that is not equal to 1 and inserting the smallest element greater than or equal to 1 that has not yet been inserted. Can another process indefinitely read the ∅ value?

First of all, notice that the first query may very well return ∅. To this end, it has to be placed before the first update in the linearization. The same happens for the second, third and so on. Every query can therefore return ∅, but can all? In that sense, the first update should be placed after an infinite number of queries. Is this allowed?

The definition of sequential consistency as defined in [LAM 79] is in fact not accurate enough to answer this question and no clear consensus emerges from within the scientific community about the most appropriate interpretation. However, both possible interpretations lead to very different criteria. In [SEZ 15], Sezguin shows that several objects including the stack and the queue can be implemented in a sequentially consistent manner without exchanging a single message if the linearization point can be deported to infinity. Conversely, in [ATT 94], Attiya and Welch show that even in a synchronous system, waiting for time proportional to the uncertainty in the latency of the network are necessary for at least one operation (push or pop). Attiya and Welch's interpretation is thus considerably stronger.

This example demonstrates the importance of accuracy in modeling shared objects. First, a precise definition makes it possible to avoid this kind of debate; except for one thing, a shared object can guarantee very strong properties or on the contrary very weak ones. Secondly, the difficulties encountered during the specification are a good indicator of the limitations of a model. In this specific case, if we can place an update after an infinite number of queries why could we not place a query after an infinite number of updates? Is a process able to read the state {1} corresponding to the set of inserted values and never deleted by the first process, although this state will never be reached? The successive application of an infinite number of updates is called a hypertask *and not all hypertasks have limits. Can it read any state, for example {0}, even if the value 0 is never inserted, since the path in the transition system will never specify a query after an infinite number of operations?*

We believe that this kind of questions is mainly a sign that the second interpretation is not adapted to formalize the intuition that one has of the consistency criterion. The strong version of the sequential consistency seems more natural. It is the one that is introduced in Figure 3.4. This is guaranteed by the linearization definition according to which any event is placed at a finite position.

Figure 2.3. *Sequential consistency and infinite histories*

SEQUENTIAL CONSISTENCY

✗ Not composable p. 42
✓ Decomposable p. 42
✗ Strong p. 142

Sequential consistency was initially defined by Lamport in [LAM 79] as: "the result of any execution is the same, as if the operations of all the processors were executed in some sequential order, and the operations of each individual processor appear in this sequence in the order specified by its program". This is, for example, the case in the history of Figure 2.1b (recalled underneath): if all the events are ordered according to the blue line, the sliding window register shifts from the state (0, 0) to the state (0, 1) during the write w(1) and from the state (0, 1) to the state (1, 2) during the write w(2), which is consistent with its sequential specification, and the same happens for the set.

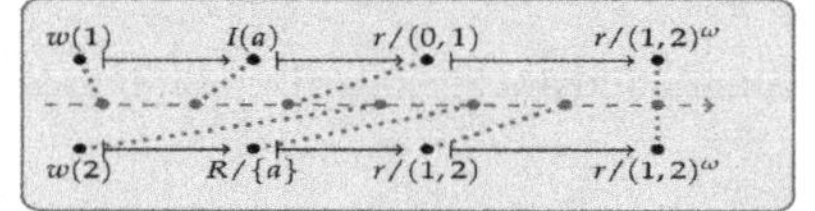

By definition a history H is sequentially consistent with regard to an abstract data type (Definition 2.1) if there is a sequence that presents the following two properties.

– *It contains the identifiers of all the events of H and the order in which they appear is compatible with the process order, that is, it belongs to* lin(H).

– *It is correct with respect to the sequential specification of the object, that is, it belongs to L(T).*

In the example above, the following linearization corresponds to these criteria:

$$w(1)/\bot \cdot I(a)/\bot \cdot r/(0,1) \cdot w(2)/\bot \cdot R/\{a\} \cdot r/(1,2)^\omega$$

DEFINITION 2.1.– Sequential consistency *is the consistency criterion:*

$$SC : \begin{cases} \mathcal{T} & \rightarrow & \mathcal{P}(\mathcal{H}) \\ T & \mapsto & \{H \in \mathcal{H} : \text{lin}(H) \cap L(T) \neq \varnothing\} \end{cases}$$

The other histories of Figure 2.1 are not sequentially consistent. Any attempt to order all the events of a history that is not sequentially consistent usually results in a cycle. A good example is the history of Figure 2.1g : a linearization that places the write w(1) before the read r/(0, 2) does not respect the sequential specification, therefore a linearization must place r/ (0, 2) before w(1) - and similarly r/(0, 1) before w(2)– which is impossible without violating the process order. Here, it can be seen what makes sequential consistency difficult to implement: the two processes must necessarily synchronize to break this cycle.

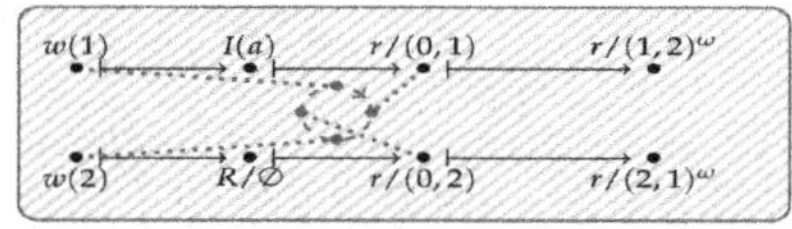

Notice that the history of Figure 2.1c (on the right) is not sequentially consistent as well. It turns out that even though it is possible to establish a total order on all the events and that once ordered, the sequential specification is satisfied (because any finite prefix satisfies it), this total order does not describe a valid linearization because the events of the second process have an infinite number of predecessors in the total order.

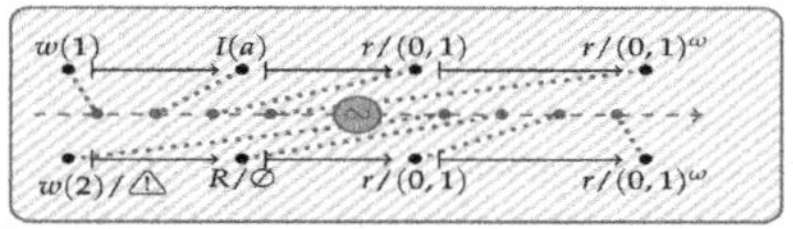

Figure 2.4. *Sequential consistency*

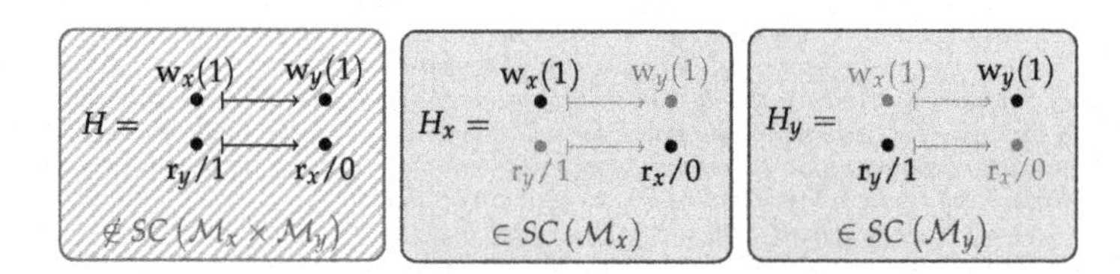

Figure 2.5. *Sequential consistency is not composable. For the color version of this figure, see www.iste.co.uk/perrin/distributed.zip*

Sequential consistency is not composable, as shown in Figure 2.5. Two processes share a memory instance composed of two registers x and y. The first process writes 1 in x and then 1 in y. The second process reads 1 in y then 0 in x. If we study each register independently, the two sub histories are sequentially consistent: the read of x

is made prior to its write and the read of y follows its write. Yet, the union of these two orders and of the process order contains a cycle and therefore the complete history is not sequentially consistent.

The original definition of sequential consistency by Lamport [LAM 79] can be applied to any abstract data type. In Figure 2.4, we simply transcribe this definition in our model.

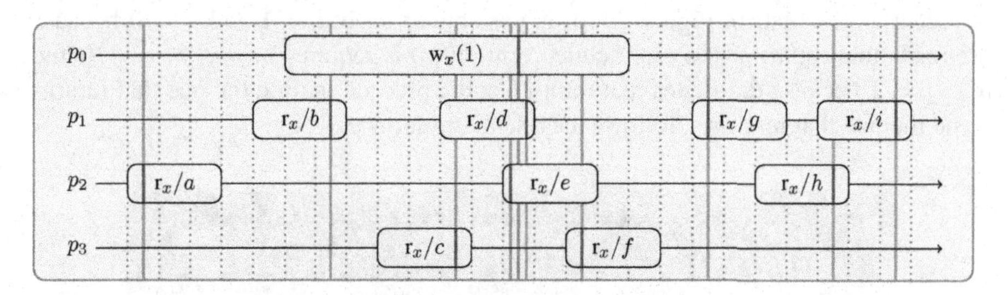

Figure 2.6. *A register shared between four processes.*

2.2.2. *Linearizability*

In 1986, Lamport proposed three specifications for the shared register: the safe register, the regular register and the atomic register [LAM 86]. The peculiarity of his model is that concurrency is defined with reference to real time. More specifically, each event has a *beginning* located at the instant when the operation is called and an *end* located at the instant when the operation terminates. Two operations are concurrent if their performances overlap in time, i.e. if the end of one is not located before the beginning of the other. Lamport further simplified the problem by assuming that only one process is allowed to write to the register, while others can only read. This allows a natural order to be established on the writes and therefore enables us to discern when a write is more recent than another. Reads can always be concurrent with writes or with other reads. Figure 2.6 presents a concurrent history in which a register is shared between four processes. Only process p_0 can write to the register.

Safe register. Safety guarantees that if a read is not concurrent with any write, then the read value must be the last written value, or the initial value if no write has happened yet. In Figure 2.6, $a = 0$ and $g = h = i = 1$. Reads concurrent with a write are not specified: they can return any value accepted by the type stored in the register. The example that best illustrates the safe register is the split-flap display showing departures in a number of train stations: updates do take a while and if one takes a photograph during editing, it is possible to see completely inconsistent information. In Figure 2.7, the cities visited by the 12:55 Thalys after Brussels are unreadable.

Regular register. In addition to safety, regularity requires that reads concurrent with a write either return the last value whose write is complete or the value being written. In Figure 2.6, b, c, d, e and f are all either 0 or 1. On the other hand, there is no requirement for successive reads of the same process to see increasingly more recent values. For example, it is possible to get $b = 1$ and $d = 0$.

Atomic register. Atomicity overcomes this problem by imposing that if two reads are not concurrent, then the second must compulsorily return a value at least as recent as the first. In Figure 2.6, it is possible to have $d = 1$ and $e = 0$ because the reads that return d and e are concurrent, but this requires $b = c = e = 0$ and $d = f = 1$ because the writes that return b and c precede in time the one that returns e and the one that returns f follows the one that returns d.

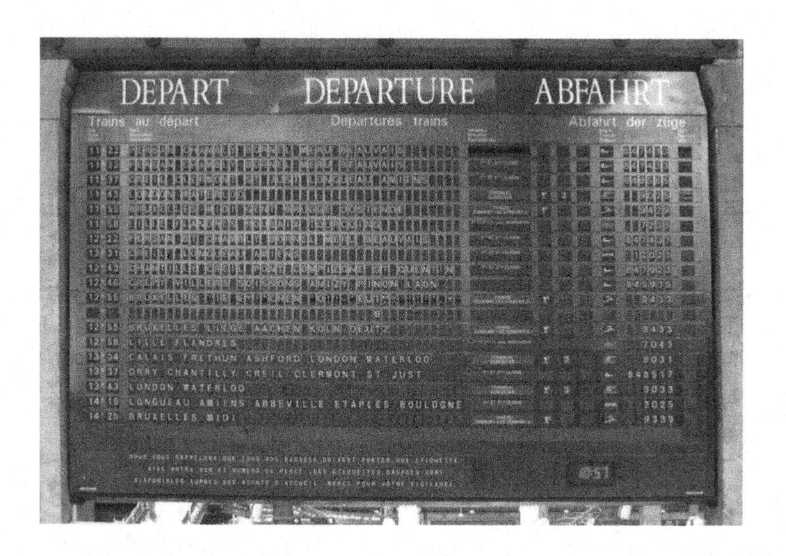

Figure 2.7. *Old departures display, in Gare du Nord, Paris [CAR 05]*

Four years later, in 1990, atomicity has been extended to other abstract data types by Herlihy [HER 90] using the name *linearizability*. A history H is linearizable for an abstract data type T if, for any event of H, there is a *linearization point* in time located between the beginning and the end of the event such that the sequential history obtained by projecting each event on its linearization point is correct with respect to the sequential specification of T. The difference with sequential consistency is real time: in linearizability, if the return of an event e temporally precedes the call to the operation of an event e', then e must be placed before e' in the linearization. For example, the history given in Figure 2.2(a) is not linearizable because the write of the 2 precedes the read $(0, 1)$ in time but not in the linearization. Linearizability can also be understood by picturing a shared copy managed by a server, as in Figure 2.2(b). In contrast to sequential consistency and to implement linearizability, the processes send a message to the server and wait for its response for any operation and not just for the queries.

Unlike sequential consistency, linearizability is local (composable and decomposable). For this reason, it has received much more attention than sequential consistency.

Our modeling of concurrent histories does not take real time into account. Therefore, it is not possible to directly transcribe linearizability. In fact, it is possible to take real time into account starting as early as when modeling the system in the form of concurrent histories, as shown in Figure 2.8: the events are the calls to the operations and an event e precedes an event e' in the process order if the end of e temporally precedes the beginning of e'. In this model, the maximal chains according to the process order do not correspond to the set of events of a process. On the other hand, the process order is an interval order, which among other things means that for any quadruplet of events (e_1, e_2, e_3, e_4), if $e_1 \mapsto e_2$ and $e_3 \mapsto e_4$ then $e_1 \mapsto e_4$ or $e_2 \mapsto e_3$. The restriction of sequential consistency to the concurrent histories that verify this property is composable.

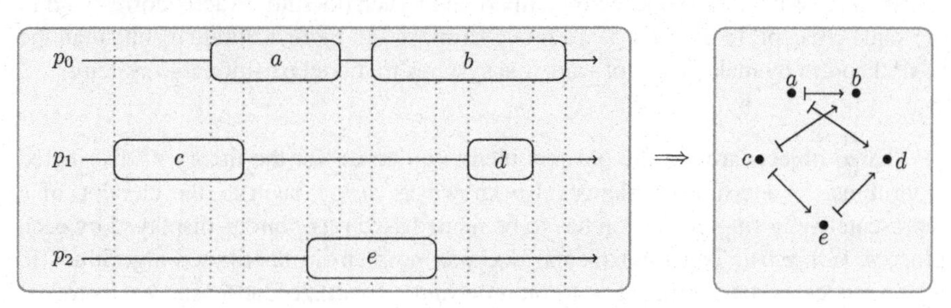

Figure 2.8. *Modeling real time in concurrent histories.*

2.2.3. *Strong consistency computability*

One of the central issues of distributed algorithms can be formulated as follows: what are the minimal assumptions necessary to implement a shared object. Attiya, Bar-Noy and Dolev have proposed an implementation of the atomic register [ATT 95] in $AS_n\left[t < \frac{n}{2}\right]$, the asynchronous distributed system using message communicating processes where strictly less than half of the processes can crash (see section 1.3.2).

This hypothesis is necessary. Lipton and Sandberg [LIP 88] have shown that to implement a sequentially consistent memory in a synchronous system in which there is a bound δ on the message transmission time, it is necessary to wait a period δ for queries or updates. Attiya and Welch have further developed this result [ATT 94] in the linearizability case. The two authors have shown that to implement atomic registers, the wait is necessary for both types of operations. Therefore, strongly consistent memory cannot be implemented in wait-free systems. Note that the intuition with a server that we gave to explain sequential consistency and linearizability (Figure 2.3) is compatible with this result: for sequential consistency, we expected a response from the server during queries but not during updates, yet the wait was necessary for all operations in linearizability. The same result is shown for the queue.

This result is very similar to the CAP theorem, originally presented as a conjecture by Brewer [BRE 00], then formalized and proved by Gilbert and Lynch [GIL 02]. It is impossible to implement strong consistency (C for strong consistency) guaranteeing that all operations complete (A for availability) in a system subject to partitioning (P for partition tolerance). However, in large-scale networks, events such as partitions regularly occur in practice [FEK 01, VOG 09]. This aspect is nonetheless subject to criticism [KLE 15]; because the notion of a partitionable system has never been precisely defined and can prove to be confusing. The model based on message loss proposed by Gilbert and Lynch does not exactly correspond to the same concept. In Chapter 5, we prove stronger and more accurate results than the CAP theorem by making use of wait-free systems to model partitionable systems.

Shared objects are not the only problems of interest for the theory of distributed algorithms. In decision problems (also known as *tasks*), such as the election of a representative, a single decision has to be made based on opinions displayed by each process. *Consensus* is a fundamental decision problem in distributed algorithms in which processes strive to agree on a common value. To utilize Consensus, each process p_i *proposes* a value v_i by using the operation propose(v_i) that verifies the following three properties:

Termination. If p_i is correct, it eventually *decides* the value;

Validity. Any value decided by a process has been proposed;

Agreement. There are no two processes that decide different values.

Consensus is universal: it is possible to implement the state machine, a universal model encoding all linearizable objects proposed by Lamport [LAM 84], by using the operation propose(v_i) as only communication primitive [SCH 90]. In [FIS 85], Fischer, Lynch and Paterson have showed that it was impossible to implement Consensus in an asynchronous system where at least one crash failure may occur. In [DOL 87], Dolev, Dwork and Stockmayer extended this result to numerous systems.

In [HER 91], Herlihy showed that it is not possible to implement Consensus in a distributed system composed of $n + 1$ processes, using an algorithm that implements it in a system composed of n processes, and this is true regardless of the number of algorithm instances available. Therefrom, he deduces a hierarchy to classify shared objects: the *Consensus number* of a shared object is the maximal number n that makes it possible to implement Consensus between n processes utilizing the operations of the object as the only communication primitives. An object that can solve Consensus regardless of the number of processes has an infinite Consensus number. Any shared object has a Consensus number at least equal to 1 because the Consensus with a single process is trivial. Conversely, any object that has a Consensus number greater than or equal to 2 cannot be implemented in systems where Consensus is impossible. The Consensus number of the linearizable memory is 1, that of the linearizable ADT queue is 2, whereas that of the linearizable UQ-ADT queue is infinite. For all $n \geq 1$, the memory in which it is possible to write atomically to n registers has a Consensus number of $2n - 2$.

Herlihy's hierarchy is sometimes criticized for its lack of robustness concerning composition. In [SHE 97], Shenk showed that there are objects that have a Consensus number strictly smaller than n but whose composition Consensus number is n. This result is counterintuitive because the hierarchy is precisely based on the fact that it is impossible to implement objects with a Consensus number equal to $n + 1$ using as many Consensus instances as necessary between n process.

2.2.4. *Other strong criteria*

Many other criteria have been proposed as adaptations of linearizability or of sequential consistency. Since they cannot be implemented in wait-free systems as well, we will just mention them. Quiescent consistency [DER 14] is an intermediate criterion between sequential consistency and linearizability, in which real time must be respected between two events only if they are separated by a pause period during which no operation is performed. K-atomicity [AIY 05], defined only for memory, is a weaker form of linearizability that allows a read to return a value from the last k values updated, according to the total order of linearization points; linearizability is therefore equal to 1-atomicity. Eventual linearizability [SER 10, DUB 15] weakens linearizability by accepting that linearizability is only guaranteed from a time t unknown in advance. Finally, set-linearizability [NEI 94] and interval-linearizability [CAS 15] extend linearizability to data structures which have no sequential specifications, with the objective of unifying the specification of persistent objects such as abstract data types that we consider here and single use decision-making objects such as Consensus. To this end, the authors proposed to model shared objects using a model very similar to order automata [HAA 07] which allows a set of operations to be concurrent.

2.3. Transactional systems

Generic algorithms used to implement strong consistency are very expensive, and algorithms specific to a particular object often appear to be sensitive in their designing and when it comes to proving their correctness. In parallel, transactions have emerged in two different areas: database management systems [HAE 83] and shared-memory distributed algorithms [HER 93]. In both cases, systems conventionally offer a set of basic operations for the insertion, deletion and edition of data in the case of databases and reads and writes in the case of shared memory. The *transactional systems* approach enables a sequence of basic operations to be encapsulated within *transactions*. Consistency is then provided both at the level of the basic operations and at the transaction level taken as a whole.

In databases, the expected properties of transactions are conventionally referred to by the acronym ACID [HAE 83] for *Atomicity* - a transaction can only be entirely committed or completely aborted -, *Consistency* - the transactions executed in a correct state drive the system to a correct state -, *Isolation* - transactions do not interfere with each other - and *Durability* - an accepted transaction is not questioned. Several consistency criteria have been proposed to unify the transaction specification in software transactional memory systems and in databases.

2.3.1. *Serializability*

Serializability [PAP 79] is the most often required property for transaction systems. It requires that the history, consisting of the same events as a correct sequential history, in which the events of a process occur in the same order as in this process. In other words, the concurrent history in which we have removed the aborted transactions must be sequentially consistent.

One way to implement a shared object in a transaction system is to encapsulate all the operations of the object within a transaction. What we define here as *serializability* (*Ser*) (Figure 2.9) is the consistency criterion that is achieved for the objects implemented in this manner. Serializability is very similar to sequential consistency with the only difference that events may *abort*. An aborted event is not part of the final linearization. On the other hand, the process that has called the operation is notified about the abortion. Serializability is thus weaker than sequential consistency, which can be seen as a kind of serializability in which no event may abort. For the purposes of our study, we strengthen serializability by imposing that pure queries cannot abort.

SERIALIZABILITY

✗ Not composable p. 42
✓ Decomposable p. 42
✓ Weak p. 142

Serializability [BER 87] is a criterion very similar to sequential consistency (see page55), the only difference being that some events are allowed toabort. The aborted events are not taken into account in the final linearization. On the other hand, the calling process receives the information that its call has failed, because it returns a particular value ⚠. This criterion is reinforced by prohibiting pure queries to abort.

The history of Figure 2.1c (recalled on the right) is serializable: if we remove the write w(2) which has aborted, the rest of the history is sequentially consistent.

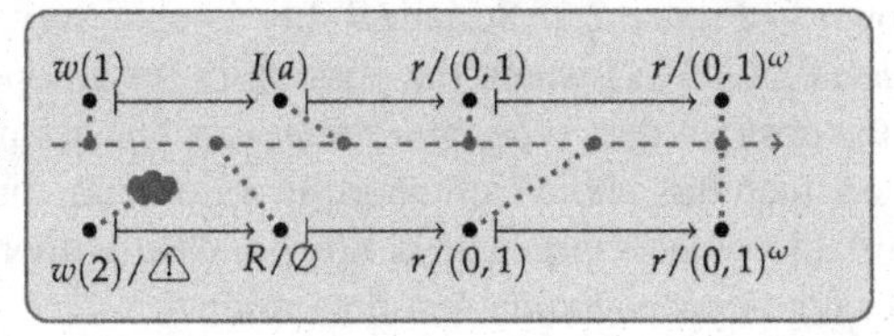

Formally, a history H is serializable for an abstract data type T (Definition 2.2). If there is a set of events $C \subset E_H$ (C for commit, from the name of the operation called at the end of the transactions in transactional systems) such that:

- *no pure query is aborted, i.e. $\hat{Q}_{T,H} \subset C$;*
- *the history $H[C/C]$ composed of the events not aborted is sequentially consistent ;*
- *any aborted event $e \in E_H \setminus C$ returns the symbol ⚠, which means that its identifier $\Lambda(e)$ is an element of $A \times \{⚠\}$ (A is the set of the input symbols of T).*

In the example above, C contains all the events except the one labeled w(2)/⚠ and the following linearization of $H[C/C]$ satisfies the sequential specification:

$$w(1)/\bot \cdot R/\emptyset \cdot I(a)/\bot \cdot r/(0,1)^\omega$$

DEFINITION 2.2.– Serializability *is the consistency criterion:*

$$Ser: \begin{cases} \mathcal{T} \rightarrow & \mathcal{P}(\mathcal{H}) \\ T \mapsto & \left\{ H \in \mathcal{H} : \begin{array}{l} \exists C \subset E_H, \ \hat{Q}_{T,H} \subset C \\ \wedge \ \text{lin}(H[C/C]) \cap L(T) \neq \emptyset \\ \wedge \ \forall e \in E_H \setminus C, \Lambda(e) \in A \times \{⚠\} \end{array} \right\} \end{cases}$$

Apart from the sequentially consistent history of Figure 2.1b, no other history of Figure 2.1 is serializable since none of their events abort and they are not sequentially consistent. This is for example the case with the history of Figure 2.1d (on the right),

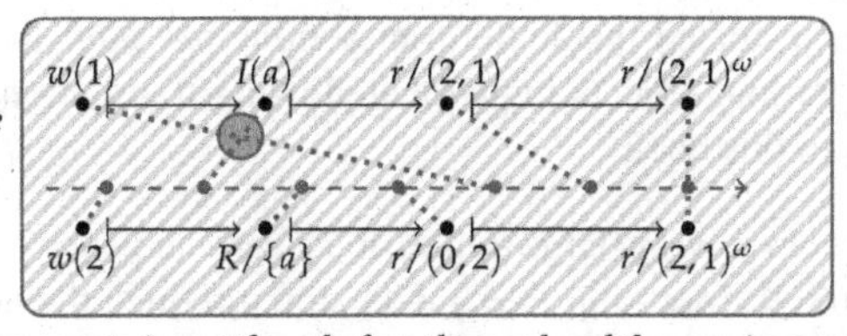

in which a cycle is formed by the requirement to insert the a before the read and the requirement to read the 1 after it has been written. On the other hand, if the two writes of the first process had been made in reverse order, the story would have been serializable.

Figure 2.9. *Serializability*

It should be noted that this definition of serializability is wider than the context of transactional systems: this consistency criterion is adapted for modeling the observed behavior with Hangouts in the experiment of the introduction (Figure 1.1). Bob's messages failed to be sent, but the issuer has been notified of the failure. This definition of serializability thus models a large number of client-server services on the Internet where an error message can be obtained when the connection is not established. The total order on the other events is implemented by the server, as shown in Figure 2.2. Serializability is also very similar to the notion of *abortable objects* [AGU 07] where the possibility for an operation to abort is directly encoded in the abstract data type rather than in a consistency criterion. However, it should be noted that the choice of aborting or not an operation is non-deterministic. The abortable object approach is not possible without adaptation in our model, which does not allow encoding non-determinism.

2.3.2. *Other criteria*

The other consistency criteria proposed to specify transactional systems are variations of serializability. *Strict serializability* [PAP 79] is to serializability what linearizability is to sequential consistency: it adds to serializability the requirement that if a transaction T_1 completes before a transaction T_2 begins, then T_1 will be placed before T_2 in the equivalent sequential execution.

The purpose of *one-copy serializability* [BER 83, KEM 09], *global atomicity* [WEI 89], *recoverability* [HAD 88], *rigorous scheduling* [BRE 91], *virtual world consistency* [IMB 12] and *opacity* [GUE 08] is to jointly specify the transactions and the basic operations offered by the transactional system (reads and writes to memory as well as insertions and deletions in databases). These considerations are rather related to the nature of the semantics of the language of the transaction specification. They provide information on the sequence of objects encapsulated inside the transactions but offer the same consistency as serializability regarding objects implemented using transactions, a perspective that is of interest to us in this book.

On a very large scale, it is very expensive to guarantee the ACID properties. NoSQL (not only SQL) data stores reduce the consistency they offer for the benefit of performance. Several recent works [KOB 10, BUR 12, XIE 14] propose to guarantee a weaker criterion to improve efficiency: eventual consistency.

2.4. Eventual consistency

2.4.1. *Eventual consistency*

Introduced in the 1990s [TER 95, PET 97] and popularized by Vogels in [VOG 09], eventual consistency (EC) is especially related to replicated objects [SAI 05], of which every process locally maintains a local copy of the object, called *replica*. Eventual consistency requires that if all processes stop updating and wait long enough, all replicas will end up in the same state, i.e. the eventual convergence state. The time required to reach eventual convergence after that updates have stopped is not specified and processes are not notified when they reach this state. Eventual consistency is formally defined in our model in Figure 2.10.

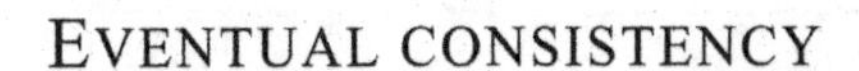

EVENTUAL CONSISTENCY

✓ Composable p. 42
✗ Not decomposable p. 42
✓ Weak p. 142

Eventual consistency is one of the few criteria that has not been initially defined for memory [VOG 09]. It requires that if all parties stop updating, all replicas end up by converging to a common state. For example, in the history of Figure 2.1e (recalled opposite), after some time, all the queries return the same value.

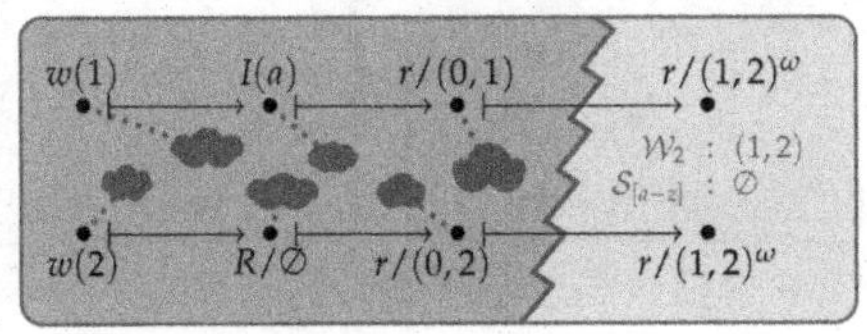

Formally, a history H is eventually consistent for an abstract data type T (Definition 2.3) if it corresponds to one of the following two cases.

– H contains an infinite number of updates (i.e. $|U_{T,H}| = \infty$). In this case, at least one participant perpetually updates, therefore the values returned by the queries can continue to develop according to the updates.

– There is a state ζ in T (the convergence state) which can explain all the queries made from a certain time (the convergence time). What characterizes such a state is the fact that the set of the events occurring after convergence E' is cofinite ($|E_H \setminus E'| < \infty$) and for any event $e \in E'$, the operation $\Lambda(e)$ can be carried out in the state ζ, thus $\zeta \in \delta_T^{-1}(\Lambda(e))$.

In the example above, E' is the set of labeled events r/(1, 2) and one of the possible convergence states is $\zeta = ((1,2), \emptyset)$. In fact, any value for the set gives a possible convergence state.

DEFINITION 2.3.– Eventual consistency *is the consistency criterion:*

$$EC : \begin{cases} \mathcal{T} & \to \quad \mathcal{P}(\mathcal{H}) \\ T & \mapsto \quad \left\{ H \in \mathcal{H} : \begin{array}{l} |U_{T,H}| = \infty \\ \vee \ \exists E' \subset E_H, \ |E_H \setminus E'| < \infty \\ \qquad \wedge \ \bigcap_{e \in E'} \delta_T^{-1}(\Lambda(e)) \neq \emptyset \end{array} \right\} \end{cases}$$

The history of Figure 2.1g (recalled opposite) is not eventually consistent as it contains only three writes (labeled w(1)/ ⊥, w(2)/ ⊥ and I(a)/ ⊥) but, for an infinite number of reads, the sliding window register must be in the state (1, 2) and for an another infinite number of reads, the state must be (2, 1).

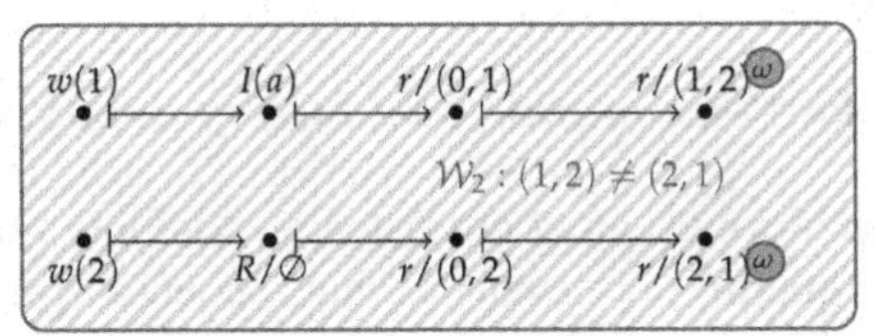

Figure 2.10. *Eventual consistency*

Eventually consistent replicated objects, currently deployed on a large scale have mainly two applications: key-value data stores such as Dynamo [DEC 07] and Cassandra [LAK 10], on the other hand, collaborative editors such as Woot [OST 06], Logoot [WEI 09], TreeDoc [PRE 09] and LSEQ [NÉD 13]. It should be observed that eventual consistency is the consistency criterion that models Skype's observed behavior in the introduction experiment (Figure I.3): both parties have not received the messages in the same order, but the display order has been resorted so that they finally see them in the same order.

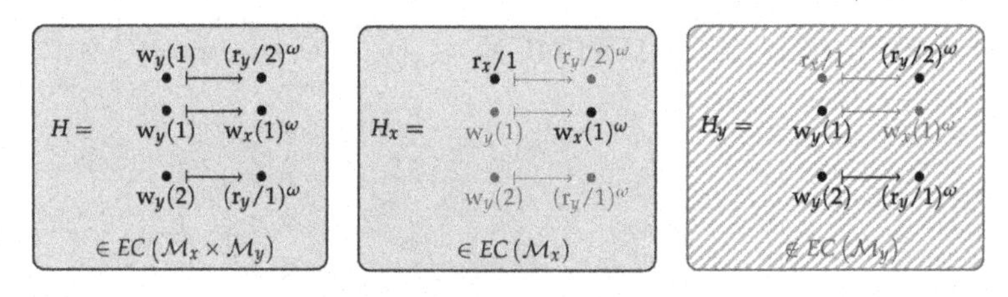

Figure 2.11. *Eventual consistency is not decomposable. For the color version of this figure, see www.iste.co.uk/perrin/distributed.zip*

Eventual consistency is composable but not decomposable. In Figure 2.11, three processes share a memory consisting of two registers. Only the events repeated an infinite number of times are important here: writes to the register x and contradictory reads of 1 and 2 to y. This history is eventually consistent only because it contains an infinite number of writes to x, but if we only consider the history concerning the register y, eventual consistency is never attained. In general, at least one of the objects of a composition is eventually consistent, but all are not necessarily so.

In addition, because of the restriction that prevents pure queries to abort in serializability, it is also stronger than eventual consistency. Indeed, in a history having a finite number of updates and an infinity of queries, almost all of them will be made in the state obtained by the linearization imposed by serializability on the updates which have not aborted.

2.4.2. *Implementation and strong eventual consistency*

To achieve eventual consistency, conflicts can be solved either during updates or during queries or even asynchronously. Strong eventual consistency (SEC) [SHA 11b] requires that they be solved during updates: if two replicas have received the same messages, they have to be in the same state. The problem with this definition is that the notions of replica and message reception are inherent to the implementation, and not available to the programmer who uses the object. Therefore, they should not appear as such in the specification. A visibility relation is introduced [BUR 14] to model the notion of message reception. Visibility is not an order relation because it is not necessarily transitive. As its name implies, strong eventual consistency, defined in Figure 2.12, is stronger than eventual consistency.

STRONG EVENTUAL CONSISTENCY

✓ Composable p. 42
✗ Not decomposable p. 42
✓ Weak p. 142

Strong eventual consistency strengthens eventual consistency by imposing on processes to converge as soon as they have received the messages corresponding to the same updates. In the history of Figure 2.1e (recalled opposite), the operations made by processes that have received the same messages are represented inside the same light frame.

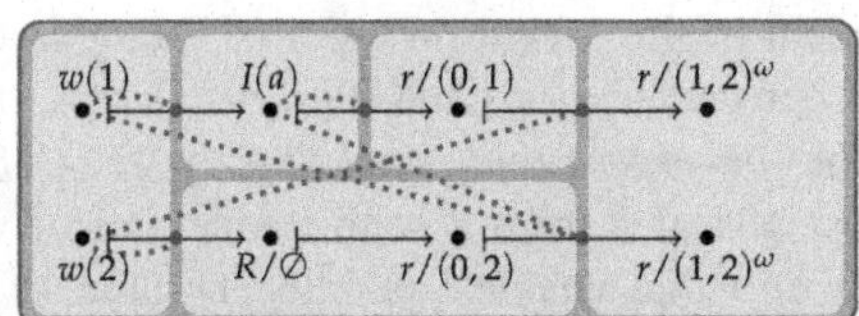

Formally, a history H is strongly eventually consistent for an abstract data type T (Definition 2.5) if there is a visibility relation $\xrightarrow{VIS}$ *(which models the reception of messages, Definition 2.4) such that the state queried during an event e is entirely determined by the set* $U_{T,H} \cap \lfloor e \rfloor_{\xrightarrow{VIS}} \setminus \{e\}$ *of updates visible by e, and a function f that describes in what state the queries must be made depending on its visible updates. We recall that* $\delta_T^{-1}(\Lambda(e))$ *is the set of states in which one event e can take place (Definition 1.4).*

In the example above, a possible visibility relation is drawn in dashes. One of the ways of defining function f is to say that the set always remains empty and that the state of the sliding window register is the sequence of the k smallest written values sorted by increasing order on integers. It turns out that no relation is imposed between the sequential specification and function f.

DEFINITION 2.4.– *Let* $H \in \mathcal{H}$ *be a concurrent history. A binary relation* $\xrightarrow{VIS} \in E_H^2$ *is a* visibility relation *of H if:*

- *it contains the process order* $\mapsto$ *of H ;*
- *it is reflexive and acyclic;*
- *any operation eventually becomes visible by all processes, that is*

$$\forall e \in E_H, \{e' \in E, e \not\xrightarrow{VIS} e'\} \text{ is finite;}$$

- *an operation visible by a process always remains visible by this same process*

$$\forall e, e', e'' \in E_H, \left(e \xrightarrow{\text{VIS}} e' \wedge e' \mapsto e''\right) \Rightarrow \left(e \xrightarrow{\text{VIS}} e''\right).$$

The set of visibility relations of H is denoted Vis(H).

DEFINITION 2.5.– Strong eventual consistency *is the consistency criterion:*

$$SEC : \begin{cases} \mathcal{T} & \to \quad \mathcal{P}(\mathcal{H}) \\ T & \mapsto \quad \left\{ H \in \mathcal{H} : \begin{array}{l} \exists \xrightarrow{\text{VIS}} \in Vis(H), \exists f : \mathcal{P}(U_{T,H}) \to Z, \\ \forall e \in E_H, f\left(U_{T,H} \cap \lfloor e \rfloor_{\xrightarrow{\text{VIS}}} \setminus \{e\}\right) \in \delta_T^{-1}(\Lambda(e)) \end{array} \right\} \end{cases}$$

The history of Figure 2.1f (recalled opposite) is not strongly eventually consistent. As a matter of fact, the first process changes twice state after the write w(1) but the history comprises only a single other write. If the read r/(1,2) has the write w(2) in its visible past, it has the same visible past as reads r/(2,1). Otherwise, it has the same visible past as the read r/(0,1). Both cases are in contradiction with the definition of strong eventual consistency.

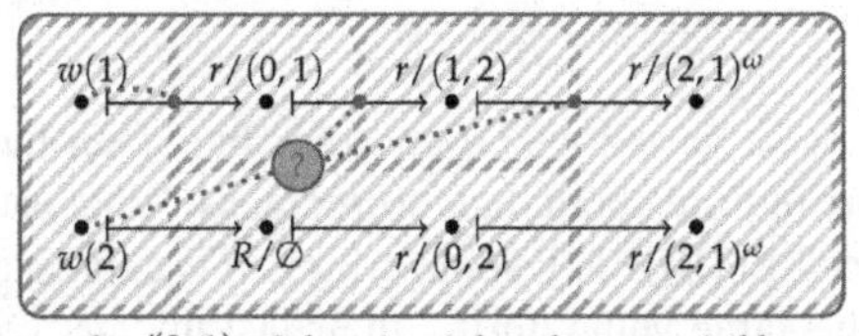

Figure 2.12. *Strong eventual consistency*

Commutative replicated data types (*CmRDT*) [SHA 11a] are abstract data types in which all update operations are pairwise commutative. For example, a counter that can be incremented or decremented is a CmRDT. The state obtained being independent of the execution order of the operations, it is easy to implement strong eventual consistency for CmRDTs. Similarly, strong eventual consistency is easy to obtain for *convergent replicated data types* (*CvRDT*), which are abstract data types whose set of states form a semi-lattice and of which all operations are increasing. The structure that enables an integer value to be proposed and which converges towards the highest proposed value is an example of *CvRDT*. In [SHA 11a], Shapiro, Preguiça, Baquero and Zawirski proved that these two classes are equal. The term *conflict-free replicated data types* (*CRDT*) represents this common class.

The same principle is used in the Bloom programming language [ALV 12], which includes a syntactic parser, CALM [ALV 11], to ensure that operations are properly commutative.

Many examples of CRDTs can be found in [SHA 11b]. On the other hand, numerous other data types which prove useful in practice are not CRDTs. For example, inserting an element in a set does not commute with the deletion of the same element. The notion of CRDT is extended to some of these objects. The shared set is one of the most studied data structures from the perspective of eventual consistency. Among all the proposed implementations of the set based on CRDTs, we can mention the *G-Set* (for grow-only set) [SHA 11b] which allows only inserting elements; the *2P-Set* (for two-phases set, also called *U-Set*) [WUU 86] which separates the inserted elements and the deleted elements into two G-Sets representing a whitelist and a blacklist and then computes the set difference during queries; the OR-Set (Observed-Remove Set, see Figure 2.13) [SHA 11b, BIE 12, MUK 14] built on the same principle but in which each element is associated with a unique identifier, which allows reinserting an element after being deleted; the C-Set (Counter Set) [ASL 11] and the LWW-Set (Last-Writer-Wins Set) [SHA 11b] that associates with each element counters whose value determines whether or not they belong to the set.

All these variations around the shared set have very different behaviors. We can therefore inquire about their correctness. What is their relationship with the set abstract data type? In [BIE 12], some criticism is expressed towards the C-Set: if all processes insert the same value during their last update, the element may still be missing after convergence. The OR-Set, which ensures that the element will be present in the same situation, would thus be more consistent with the intuition than the C-Set. However, the OR-Set shows similar inconsistencies (see Chapter 3). This raises the issue of the specification of these objects.

If eventual consistency and strong eventual consistency are not sufficient to completely specify shared objects, this is because they say nothing about the state to which the replicas must converge. Any eventual convergence state is therefore *a priori* correct. Three approaches have mainly been designed to specify the convergence state.

OR-SET

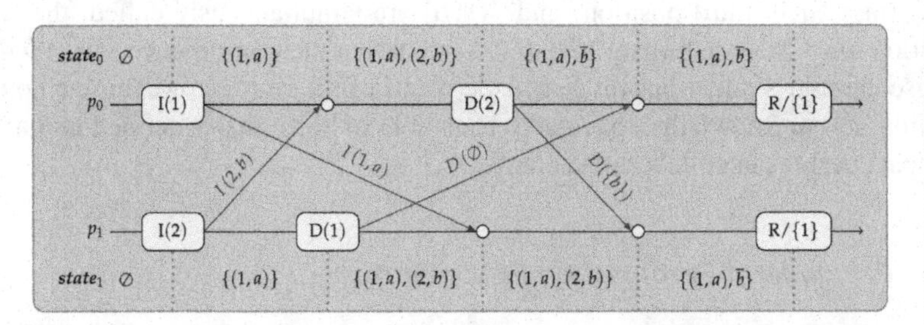

The OR-Set (for Observed-Remove Set) [SHA 11b] is one of the adaptations of CRDTs to the shared set. It has the same operations as the set abstract data type (page 29): the insert and delete operations of an element v and the query of the elements existing in the set. During an insertion, an identifier that is unique to the history is assigned to the inserted element. For example, in the above execution, identifiers a and b are assigned to both insertions. To delete an element v, the process begins by observing *the identifiers assigned to the element to be removed (hence the name of the data structure) and sends a message containing these identifiers. For example, during the delete operation of the 2, the first process observes the element(2, b) in its local replica and therefore sends the message requesting the removal of the identifier b. The element is then removed from the local replica but atombstone $\bar{b}$ marking the deleted identifier is kept to indicate that it has already been deleted. OR-Set optimizations [BIE 12, MUK 14] are intended to remove these tombstones to reduce the memory cost.*

When an insertion and deletion of the same element are concurrent (in the sense of the visibility relation), the deletion has no effect. In the above example, the insertion of thel and its deletion are concurrent and the 1 is as expected present in the set after convergence. This corresponds to the "insertion wins" strategy that is formalized by the concurrent specification of the OR-Set: the IW-Set (Definition 2.6). This specification means in the formalism of Definition 2.8 that an element v is queried during the query if there is an insertion of v (event e_i identified $I(v)/\bot$) that precedes the query according to the visibility relation but no deletion of v (event e_d identified $D(v)/\bot$) between the insertion and the query.

DEFINITION 2.6.– *Let Val be a countable set. The support IW-Set Val is the concurrent specification:*

$$\left(\mathsf{A} = \bigcup_{n \in Val} \{\mathrm{I}(n); \mathrm{D}(n); \mathrm{R}\}, \mathsf{B} = \mathcal{P}_{<\infty}(Val) \cup \{\bot\}, F\right)$$

$$F : \begin{cases} \mathcal{L}(\mathsf{A}) \times \mathsf{A} & \to \mathsf{B} \\ \left(\left(E, \Lambda, \xrightarrow{\text{VIS}}, \leq\right), \ \mathrm{I}(n)\right) & \mapsto \bot \\ \left(\left(E, \Lambda, \xrightarrow{\text{VIS}}, \leq\right), \ \mathrm{D}(n)\right) & \mapsto \bot \\ \left(\left(E, \Lambda, \xrightarrow{\text{VIS}}, \leq\right), \ \mathrm{R}\right) & \mapsto \left\{ v \in Val : \begin{array}{l} \exists e_i \in \lfloor e \rfloor_{\xrightarrow{\text{VIS}}}, \Lambda(e_i) = \mathrm{I}(v)/\bot \\ \wedge \forall e_d \in E, \\ e_i \xrightarrow{\text{VIS}} e_d \xrightarrow{\text{VIS}} e \Rightarrow \Lambda(e_d) \neq \mathrm{D}(v)/\bot \end{array} \right\} \end{cases}$$

Figure 2.13. *The OR-Set*

2.4.3. *Intention*

The notion of *intention* [SUN 98, LI 00] has initially been proposed to specify eventually consistent collaborative editors. The idea is to take into account the user's intention who has called the operations in the solution of the convergence state. For instance, if the shared document contains the word "*balad*" and that the two operations $I(l,3)$ (insert l in third position) and $I(s,6)$ are simultaneously called, the second operation must be transformed into $I(s,7)$ to obtain the convergence state "*ballads*" because the user's intention was to use the plural of the word. However, the notion of intention, to our knowledge, has never been able to be formally defined and has had no impact besides in collaborative editing.

2.4.4. *Principle of Permutation Equivalence*

The Principle of Permutation Equivalence has been proposed by [BIE 12]. It defines the compliance of an object to its sequential specification by: "If all sequential permutations of updates lead to equivalent states, then it should also hold that concurrent executions of the updates lead to equivalent states." To completely specify a convergent object, simply assigning an convergence state to each pair of non-commutative operations would then be enough.

For example, in the case of the shared set, two insertions commute. The same happens for two deletions as well as for the insertion and the deletion of two different elements. To completely specify the shared set, it would therefore suffice to define the behavior of each operation and decide what should give the concurrent execution of the insertion and of the deletion of the same element. In the case of the OR-Set, the resulting state is the one that contains the element.

The Principle of Permutation Equivalence seems to be acceptable to define the shared set because for every operation, there is at most one operation with which it does not commute. What if this is not the case? Consider the example of an abstract data type having three states ζ_a, ζ_b and ζ_c, and three operations a, b and c. Executed on its own, a leads to the state ζ_a, b to the state ζ_b and c to the state ζ_c. No pair of different operations does commute. According to the principle of permutation equivalence, it would suffice to define a behavior for each pair. One possibility is to define that $a||b$ results in the state ζ_c, $b||c$ results in the state ζ_a and $a||c$ results in the state ζ_b. However then, what should be obtained if we concurrently executed a, b and c? The three states play a symmetric role and thus none is any longer legitimate to be chosen as convergence state.

A more concrete example is the C-Set. For simple examples with only two operations, the execution is in line with expectations. The counter example that has been given in [BIE 12] involves four operations. Thus, even for the set the Principle of Permutation Equivalence is not sufficient to specify the action of concurrent operations.

2.4.5. *Concurrent specifications*

In practice, defining the behavior of the operations pairwise is not enough to completely specify an object. In the general case, an operation is not concurrent with another operation only, but with a set of operations, themselves linked together according to the visibility relation of strong eventual consistency. Burckhardt, Gotsman, Yang and Zawirski [BUR 14] proposed the notion of *concurrent specification* to specify strong eventually consistent shared objects.

Formally, the approach specifies the state returned during an operation by the *operation context* in which it is performed (definition 2.7), i.e. the set of its past operations according to the visibility relation and ordered according to the visibility relation and an *arbitration* relation, a total order that can be used to solve conflicts. A concurrent specification associates a return value to each operation according to the context in which it is called (definition 2.8).

DEFINITION 2.7.– Let A be a countable set, which plays the same role as the input alphabet in abstract data types. An *operation context* on A is a quadruplet $L = (E, \Lambda, \xrightarrow{\text{VIS}}, \leq)$ where:

– E is a finite set of *events*;

– $\Lambda : E \to \mathsf{A}$ is the *labeling function*;

– $\xrightarrow{\text{VIS}}$ is a *visibility relation* on the events of E (definition 2.4))

– $\leq$, the *arbitration* , is a total order on the events of E.

The set of operation contexts on A is denoted by $\mathcal{L}(\mathsf{A})$

DEFINITION 2.8.– A *concurrent specification* is a triplet $(\mathsf{A}, \mathsf{B}, F)$ where :

– A is a countable set called *input alphabet* ;

– B is a countable set called *output alphabet* ;

– $F : \mathcal{L}(\mathsf{A}) \times \mathsf{A} \to \mathsf{B}$ is the specification function that indicates what return symbol is expected for each input symbol in a given operation context.

DEFINITION 2.9.– Let $H = (\Sigma, E, \Lambda, \mapsto)$ be a concurrent history and $(\mathsf{A}, \mathsf{B}, F)$ a concurrent specification.

It is said that H is strongly eventually consistent with respect to $(\mathsf{A}, \mathsf{B}, F)$ if there is a visibility relation $\xrightarrow{\text{VIS}}$ and a total order $\leq$ on E_H (the arbitration) such that for any event $e \in E$, the operation that labels e is consistent with the application of the concurrent specification on the context formed by the past of e according to the visibility relation:

$$\forall e \in E, \Lambda(e) = \alpha/\beta \Leftrightarrow F((\lfloor E \rfloor_{\xrightarrow{\text{VIS}}}, \Lambda, \xrightarrow{\text{VIS}}, \leq), \alpha) = \beta$$

The concurrent specification of the OR-Set is called *IW-Set* (for Insert Wins Set), formally defined in Figure 2.13. It expresses the fact that an element x belongs to the set if and only if an insertion event of x is visible at the time of the query, but no deletion event of x is located between the insertion and the query with respect to $\xrightarrow{\text{VIS}}$.

In practice, this approach has also several limitations:

1) The first limitation is its complexity compared to sequential specifications. Abstract data types are specified by means of transition systems in which the specification of each operation is independent. Transition systems naturally describe a language: the sequential specification. On the contrary, in the concurrent specification it is important to give a specification for all possible operation contexts, which generally results in a specification as complex as the implementation itself.

This limitation raises two questions. First, what is the relationship between concurrent specifications and the intuition that one has of the data structures that they aim to specify? In short, concurrent specifications do not make it possible to justify why the OR-Set would be more consistent with the intuition of a set than the C-Set, as it is claimed in [BIE 12]. Second, the existing verification techniques are based on transition systems. What techniques could be used to formally verify concurrent specifications?

2) The second limitation is that concurrent specifications can be utilized only to specify shared objects verifying a consistency criterion at least as strong as strong eventual consistency. In effect, they are based on the notion of visibility which is one of the characteristics of strong eventual consistency. The function F is deterministic, two states with the same past according to the visibility relation are necessarily in the same state. For example, how can we specify eventually consistent objects whose conflicts are solved asynchronously according to this technique?

3) The third limitation is that the visibility relation is directly based on the flow of messages in the system. Intuitively, two events a and b are concurrent if the process that executes b has not received the message of the one that executes a at the time it starts b and vice versa. However, messages do depend on the implementation and not of the specification.

Finally, we can say that this approach is a good technique for *modeling* certain shared objects: it makes it possible to finely describe the behavior of these objects and to prove some properties about the correction and the complexity of the algorithms, as illustrated in [BUR 14]. On the other hand, it is not satisfactory as a specification technique. In Chapter 3, we will introduce a new consistency criterion, *update consistency* that uses sequential specifications to specify convergence states.

2.5. Shared memory

In addition to large-scale message-passing systems, strongly consistent memory is also too expensive in shared-memory parallel architectures used for high-performance computing. New memory models had to be designed to weaken linearizability. Several studies aim to identify them [ADV 10, ADV 96, MOS 93]. We will not focus on the attempts to classify these models according to a few basic properties such as *session guarantees* [TER 94], which leave out important subtleties of memory models.

In this part, memory is seen as a particular abstract data type and memory models will each be extended into a consistency criterion which, applied to the memory type, results in the original memory model. There are potentially many ways to extend a memory model into a consistency criterion; our choice was to retain the criterion that best corresponds to the intuition brought forward by the authors of each memory model.

2.5.1. *PRAM memory*

In [LIP 88], Lipton and Sandberg proposed *PRAM* memory (for Pipelined Random Access Memory, also called FIFO memory) which guarantees that all updates are seen by all processes and that the process order is respected. The authors proposed a specification for PRAM: a history H is PRAM if, for any process p, there is a linearization containing all the writes to the registers and the reads carried out by p such that each read of a register x returns the last value written to x in the order of the linearization if it exists or the initial value 0 otherwise.

In other words, the linearization requested for each process verifies the sequential specification of the memory. This property is thus easy to extend to other abstract data types. We formally define the corresponding consistency criterion, pipelined consistency (PC), in Figure 2.14.

Pipelined consistency is weaker than sequential consistency, whereby the order of the linearization of all process is the same. It is not comparable with serializability and eventual consistency because different processes can execute non-commutative updates in a different order, resulting in a different state.

PIPELINED CONSISTENCY

✗ Not composable p. 42
✓ Decomposable p. 42
✓ Weak p. 142

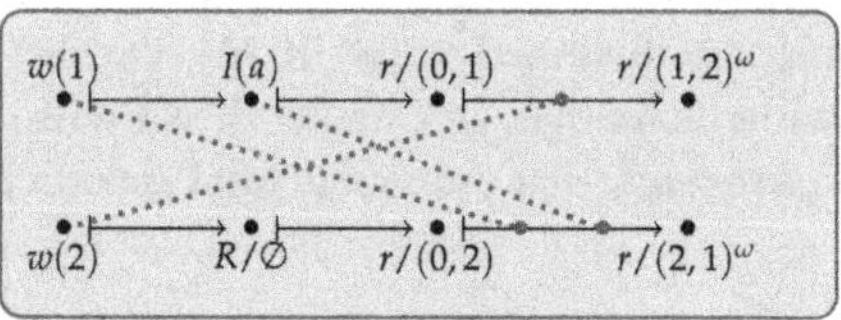

According to PRAM memory [LIP 88], there is a linearization for each process p, which contains all of the queries of p and the updates of all the processes, which satisfies the sequential memory specification. The same definition can be applied with the object of Figure 2.1g (recalled opposite): the write w(2) of the second process can be placed between the reads of the first process returning (0, 1) and (1, 2) and both writes of the first process can be placed between the reads/r(0, 2) and r/ (2, 1) of the second.

The level of formalism adopted in [LIP 88] to specify PRAM memory being satisfactory, we merely adapt the definition in our formalism. The updates correspond to the transition of the operations (symbols of the input alphabet α) and queries to their return value (symbols of the output alphabet β). The projection operator allows separating the update portion of an operation from the query portion: if H is a concurrent history and p is a process, history $H[E_H/p]$ *contains the same events as H, but the return values (queries) of the events that are not in p are hidden. Thus, a history H verifies pipelined consistency for an abstract data type T (Definition 2.10) if, for each process* $p \in P_H$*, there is a linearization of* $H[E_H/p]$ *which belongs belongs to L(T). In the above history, the following words are linearizations for both processes:*

$$w(1)/\bot \cdot I(a)/\bot \cdot r/(0,1) \cdot w(2) \cdot R \cdot r \cdot (r \cdot r/(1,2))^{\omega}$$
$$w(2)/\bot \cdot R/\emptyset \cdot r/(0,2) \cdot w(1) \cdot I(a) \cdot r \cdot (r \cdot r/(2,1))^{\omega}$$

DEFINITION 2.10.– Pipelined consistency *is the consistency criterion:*

$$PC: \begin{cases} \mathcal{T} & \rightarrow & \mathcal{P}(\mathcal{H}) \\ T & \mapsto & \{H \in \mathcal{H} : \forall p \in \mathscr{P}_H, \mathrm{lin}(H[E_H/p]) \cap L(T) \neq \emptyset\} \end{cases}$$

Pipelined consistency guarantees three properties:

Each process integrates all update operations. *The history of Figure 2.1c (on the right) does not satisfy this property: none of the processes sees the write w(2). Thus, in any linearization for the first process, the write w(2) is followed by an infinite number of reads r/(0, 1), which does not respect the sequential specification.*

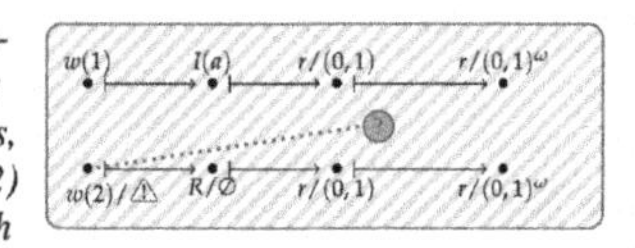

The process order between the updates is respected. The *The history of Figure 2.1d (on the right) does not satisfy this property because the second process sees the write I(a) before the read R/{a} and the write w(1) after the read r/(0, 2), therefore there is an inversion between I(a) and w(1).*

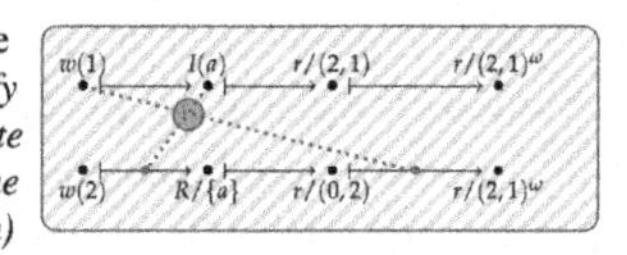

The order of the operations already integrated is never questioned. *This is due to the fact that a single linearization is required per process: for every query, the returned state is consistent with the state obtained by applying all updates that precede it in the linearization order. The history of Figure 2.1e (on the right) does not meet this property because during the integration of the write w(1), the sliding window register of the second process shifts from the state (0, 2) to the state (1, 2), which is not consistent with its sequential specification. This property is very similar to the concept of* state locality *discussed on page 146.*

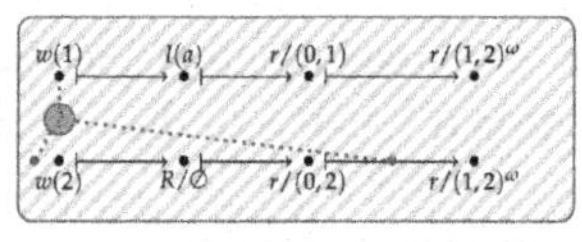

Figure 2.14. *Pipelined consistency*

Note that pipelined consistency is the consistency criterion that models the behavior observed for WhatsApp in the experiment of the introduction (Figure I.2): each party receives all messages and the fact that messages are placed at the end of the message queue upon their receiving makes it possible to guarantee that the evolution of the content of the screen of each user over time is consistent with the sequential specification of the sliding window register.

2.5.2. *Cache consistency and slow memory*

In large-scale parallel systems, access to shared memory is generally achieved by the intermediary of local caches at each process that allows very fast writes and guarantees asynchronous updating. At any moment, the latest version of each register may be in the caches of different processes. If a process reads two registers into its own cache, it may get the latest version for one of the registers but an older version for the other. The consistency criterion guaranteed by such a system is called *cache consistency* [GOO 91] (or simply coherence [GHA 90]). In [SOR 11], Sorin, Hill and Wood showed that such memory guarantees sequential consistency with each register separately. Since sequential consistency is not composable, it is not guaranteed in all of the memory. In [ADV 90], Adve listed all conditions that a program should verify to guarantee a sequentially consistent execution by utilizing memory that verifies cache consistency.

Sorin, Hill and Wood's characterization is useful to extend the criterion to other objects than registers. For memory, cache consistency is local (composable and decomposable): for a single register, cache consistency is equivalent to sequential consistency, therefore memory verifies cache consistency if, and only if, all the registers that it is made of verify it. It is also the strongest composable criterion weaker than sequential consistency since registers ought to verify sequential consistency. Cache consistency is therefore the criterion $SC^\star$ (see Figure 2.15). Cache consistency is similar to the notion of sequential consistency per object implemented in Yahoo!'s database [COO 08].

According to the definition of composition closure, no consistency criterion weaker than sequential consistency that is not weaker than cache consistency is composable. In particular, since PC is not comparable to $SC^\star$, there is no composable consistency criterion between pipelined consistency and sequential consistency.

COMMENT 2.1.– For any criterion $C \in \mathcal{C}$, if $PC \leq C \leq SC$, then C is not composable.

Slow memory [HUT 90] is built similarly to cache consistency, but pipelined consistency replaces sequential consistency. This is therefore the consistency criterion $PC^\star$.

CACHE CONSISTENCY

✓ Composable p. 42
✓ Decomposable p. 42
✗ Strong p. 142

Sequential consistency is not composable, but what happens when we use multiple sequentially consistent objects in the same program? Concerning the memory, the consistency criterion in which each register is sequentially consistent regardless of others is called cache consistency *[SOR 11]. More generally, a history H verifies cache consistency with a product of abstract data types $T_1 \times \cdots \times T_n$ (Definition 2.11) if all the sub-histories that only contain the events of a single ADT are sequentially consistent. For example, the history of Figure 2.1d (recalled here) verifies cache consistency, the following linearizations are correct for the set and for the sliding window register.*

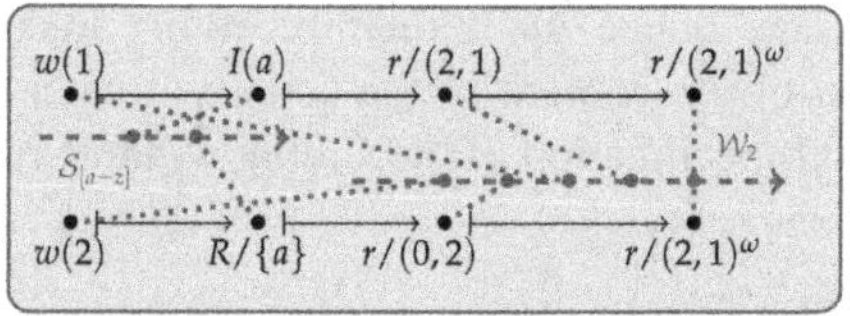

$$I(a)/\bot \cdot R/\{a\} \qquad w(2)/\bot \cdot r/(0,2) \cdot w(1)/\bot \cdot r/(2,1)^\omega$$

DEFINITION 2.11.– Cache consistency *is the composition closure SC^* of sequential consistency (see Definition 1.21).*

The history of Figure 2.1e (opposite) does not verify cache consistency because the sub-history consisting only of the events concerning the sliding window register is not sequentially consistent.

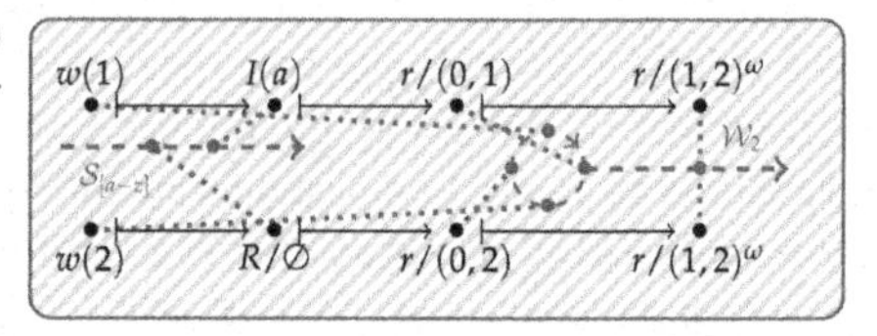

Figure 2.15. *Cache consistency*

2.5.3. *Processor consistency*

Processor consistency has initially been introduced at the same time as cache consistency [GOO 91] and then formally defined in [AHA 93]. Similarly to PRAM memory, processor consistency requires for each process p that there is a linearization containing all the writes and reads of p. In addition, two writes to the same register have to appear in the same order in the linearizations of all processes. We do not provide a formal definition of process consistency in our model, because this criterion will not be of any use to us further on.

This criterion is interesting because it shows the limitations of lattice structures to study consistency criteria. From the definition, it would be tempting to define processor consistency as the conjunction of pipelined consistency and cache consistency ($PC + SC^\star$). In [AHA 93], it is observed that processor consistency is in fact strictly stronger. Figure 2.16 illustrates this point by presenting a history verifying both pipelined consistency and cache consistency but not processor consistency. Subsequently, we will not seek to prove that a criterion C can be decomposed into the conjunction of two criteria $C_1 + C_2$, because there is generally something (the linearizations of every process in the case of processor consistency) that binds the properties of C_1 and C_2 in C.

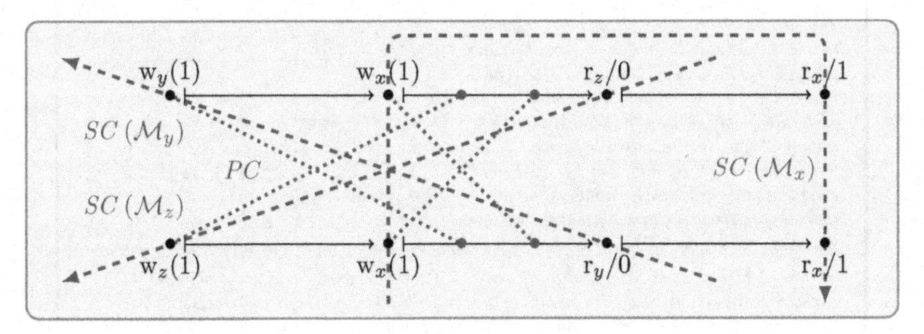

Figure 2.16. *Processor consistency versus cache consistency and PRAM*

2.5.4. *Causal memory*

Causal memory [AHA 95] strengthens PRAM memory by imposing that the order in which each process sees the updates follows a *causal order* that contains not only the process order, but also the *writes-into order*: the write of a value to a register must precede the read of this value in this register according to the causal order. Causal memory is formally defined in Figure 2.17.

Causal memory can be implemented in wait-free systems using causal reception (see section 1.3.2). An optimal algorithm is presented in [BAL 04]. Partial replication can be used to allow processes that do never access certain registers to reduce the necessary space in their local memory [HÉL 06]. Causal memory is too expensive to be implemented in very large-scale systems because the size of the control information that must necessarily be added to messages is linear regarding the number of processes [CHA 90]. A number of works [BAI 12, TSE 15] have focused on reducing the complexity by letting the application choose for which registers causal relations are important.

The definition of causal consistency independently of the abstract data type is a complicated problem. In fact, the writes-into order plays a central role in the definition of causal memory, and it is highly related to the very notion of register. Several works have aimed to expand causal consistency to other data types to conjugate it with eventual consistency. Most of them [SUN 98, LLO 11, FAR 06] simply assume that the implementation makes use of causal reception. This approach is not satisfactory because it is completely dependent on the system. In [ATT 15], Attiya, Ellen and Morrison defined causal consistency as strong eventual consistency in which the visibility relation is transitive. Nevertheless, such a criterion applied to memory does not result in causal memory that does not impose eventual consistency. We will explore this issue in more detail in Chapter 4.

CAUSAL MEMORY

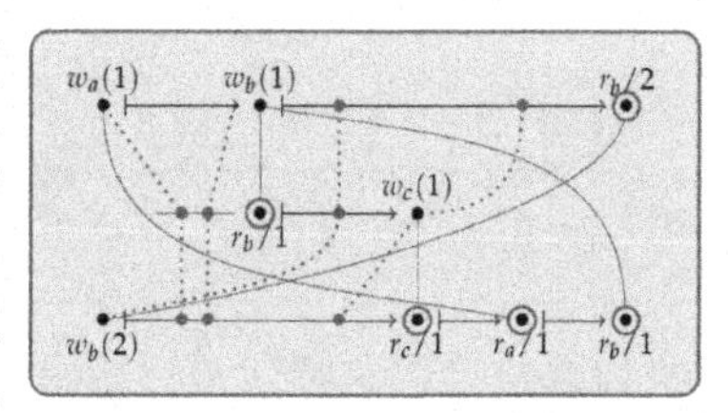

The definition of causal memory (Definition 2.13) explicitly builds a causal order composed of the process order and of a writes-into order $\multimap$ *that relates each read to the update that has written the read value (Definition 2.12). In the example opposite, the writes-into order is indicated by solid lines completed by circles surrounding the read. If the union of the process order and the writes-into order is acyclic, a causal order* $\xrightarrow{CM}$ *containing it can be built. Let X be a set of register names. A history H is* M_X*-causal if, as with PRAM memory, it is possible to build, for each process p, a linearization containing the reads of p and all the writes of* E_H *which satisfy the causal order and the sequential specification of memory, that is to say a linearization of* $\text{lin}\left(H \xrightarrow{CM} [E_H/p]\right) \cap L(M_X)$. *In the example, the three linearizations are modeled by the successions of blue and black points for all three processes.*

DEFINITION 2.12.– *Let a countable set of register names be X and a concurrent history be* $H = (\Sigma, E, \Lambda, \mapsto)$. *A relation* $\multimap$ *is a* writes-into order *if the following three properties are met.*

– *A writes-into order binds the writes to the reads of the same value to the same register (for instance* $w_a(1) \multimap r_a/1$*):*

$$\forall e, e' \in E_H, \exists x \in X, \exists n \in N, \; e \multimap e' \Rightarrow \Lambda(e) = w_x(n) \wedge \Lambda(e') = r_x/n$$

– *A read can only be related to a single write:*

$$\forall e \in E_H : /\lfloor e \rfloor_{\multimap}/ \leq 1$$

– *A read that is not related to a write returns the initial value 0:*

$$\forall e \in E_H, \exists x \in X, \exists n \in N, \; (\Lambda(e) = r_x/n \wedge \lfloor e \rfloor_{\multimap} = \emptyset) \implies n = 0$$

DEFINITION 2.13.– *Let X be a countable set of register names. A concurrent history H is* $\mathcal{M}_X$-causal *if there is a writes-into order* $\multimap$ *such that:*

– *there is a partial order* $\xrightarrow{CM}$ *that contains* $\multimap$ *and* $\mapsto$.

– $\forall p \in P_H, \text{lin}\left(H \xrightarrow{CM} [E_H, p]\right) \cap L(M_X) \neq \emptyset$.

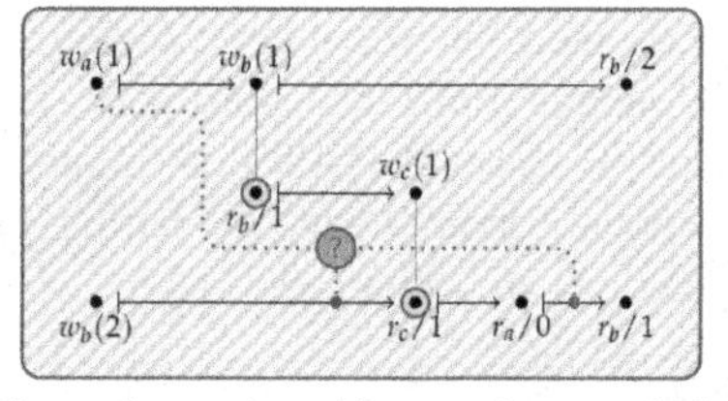

The history opposite is not $M_{[a-z]}$*-causal. In fact, since the first read of the second process returns a non-zero value, it must have a predecessor according to the writes-into order. This history can only be the event labeled* $w_b(1)$, *the only write to b. Similarly, the write* $w_c(1)$ *precedes the read* $r_c/1$. *Thus, for the linearization requested for the third process to satisfy causal order, the write* $w_a(1)$ *must precede the read* $r_a/0$ *and no write* $w_a(0)$ *may be interleaved between the two, which goes against the sequential specification.*

Figure 2.17. *Causal memory*

2.5.5. *Other memory models*

In *real-time causal consistency* [MAH 11], causal order is defined in the same way as in causal memory, with the additional constraint that a write cannot precede a read in causal order and succeed it in real time. The causal order thus obtained is then used in the same way as the visibility relation in strong eventual consistency. The authors proved that this criterion can be implemented in wait-free systems and that no strictly stronger criterion can therein be implemented. However, the proof only considers the implementations in which exactly one message is broadcast by write and none by read, which limits the scope of this result.

In peer-to-peer networks, it would be very expensive to guarantee that all processes will receive a broadcast message. It is therefore possible that after a write, those that have received the value leave the system and that the written value be lost. Weakly persistent causal memory [BAL 06, MIL 06] is a weakened version of causal memory that only requires the linearizations of the processes to contain the values that have been written an infinite number of times.

Many memory models are built by hybridizing several of the previous models, each criterion being guaranteed in different situations. In *weak consistency* [DUB 86], all registers do not play the same role. Some registers are sequentially consistent synchronization variables. The sequence of operations on these variables establishes a framework in which the operations on the other variables are included. Let x and y be synchronization variables and z any register utilized by two processes p_1 and p_2. If p_1 writes to z then accesses x while p_2 accesses y then reads z, and the access to x precedes that to y in the linearization, then the read to z by p_2 will return a value at least as recent as that written by p_1. Released consistency [GHA 90] and entry consistency [BER 91] are variants of weak consistency.

Fish-eye consistency [FRI 15] is based on a *proximity relation* between the nodes of a network. For example, the main services of cloud computing employ several data centers around the world. Computers in a same data center can communicate more effectively with each other than computers in different centers. They are thus related by the proximity relation. Fish-eye consistency guarantees sequential consistency between neighboring nodes and causal memory between remote nodes. This idea is similar to the algorithm proposed in [JIM 08] which automatically adapts to any problems that may occur in the system: during normal periods, sequential consistency is implemented, but in the case of partition, a minimal service is ensured to guarantee the causal memory.

2.6. Conclusion

In this chapter, we have studied the concepts of consistency proposed by different communities. To this purpose, we have expressed them in the model of Chapter 1. Among all these concepts, four seem to us particularly important: sequential consistency, serializability, eventual consistency and pipelined consistency. It should be noted that our model allows a better system abstraction than the one for which the criteria have initially been proposed. For instance, our definition of serializability applies equally to both transactional systems and to accessing Internet sites by considering temporary disconnections. Nevertheless, two gray areas still remain to be clarified:

Specification of eventually consistent objects. Eventual consistency has a very different definition from the other criteria presented in this chapter. In fact, the sequential specification $L(T)$ is not used at all in its definition, which makes this criterion extremely weak. It is this particularity that has made concurrent specifications necessary. Update consistency described in Chapter 3 takes into account the sequential specification, which greatly strengthens eventual consistency and responds to the specification issue of the convergence state.

Causal consistency. The definition of causal memory contains in itself the definition of the relations between reads and writes. The notion of writes-into order has no natural expression in our formalism. For this reason, extending causal consistency to all abstract data types is a difficult problem. We will explore this issue in more details in Chapter 4.

The history of linearizability consolidates our approach consisting of defining consistency criteria regardless of the abstract data types. Similarly to many weak criteria, it was initially only a specification of shared memory: the atomic register. Its extension to other abstract data types has turned it into a central tool in the study of distributed systems, in particular with regard to computability in distributed systems. We are studying the impact of our approach on the study of computability in wait-free systems in Chapter 5.

The question of composability raises a problem. In effect, we have seen that very few criteria were naturally composable and composable criteria on the path to sequential consistency are in fact very weak. Another approach is therefore required to develop applications using multiple objects.

3

Update Consistency

3.1. Preamble

None of the three scenarios featured in the introduction exactly describes what happened that day between Alice and Bob. In truth, they used neither Hangouts nor WhatsApp, nor even Skype but Facebook Messenger[1] (their conversation is transcribed in Figure 3.1). It all started as we thought in the case of WhatsApp: Bob did not notice that Alice had interpreted his "Obviously" as an answer to "Are you still angry?" and not to "Coffee break?", and later he needed all his persuasive power to undo all that had happened with Alice. Yet, when they talked about it in the evening, Bob was unable to justify himself: the order of the messages had changed since the afternoon, in such a way that his message thread was now the same as Alice's.

The joy of meeting each other had promptly erased everyone's concerns, especially as their misadventure reminded them of their mutual friend Carole's prank. The previous April 1st she presented them with her own instant messaging application that randomly replaced sent messages. Alice had also invited Bob to have coffee who had answered by his usual "Obviously", but the message "Certainly not:(" was transmitted. Fortunately, this time no consequence occured, Bob was able to see the transformation of the message (the images that appear in the transcript of their conversation in Figure 3.2 fortunately do not originate from an actual experiment[2]).

1 https://www.messenger.com/

2 Such a modification of the messages can be made possible due to security flaws such as the one that Facebook suffered in 2016 [TEA 16].

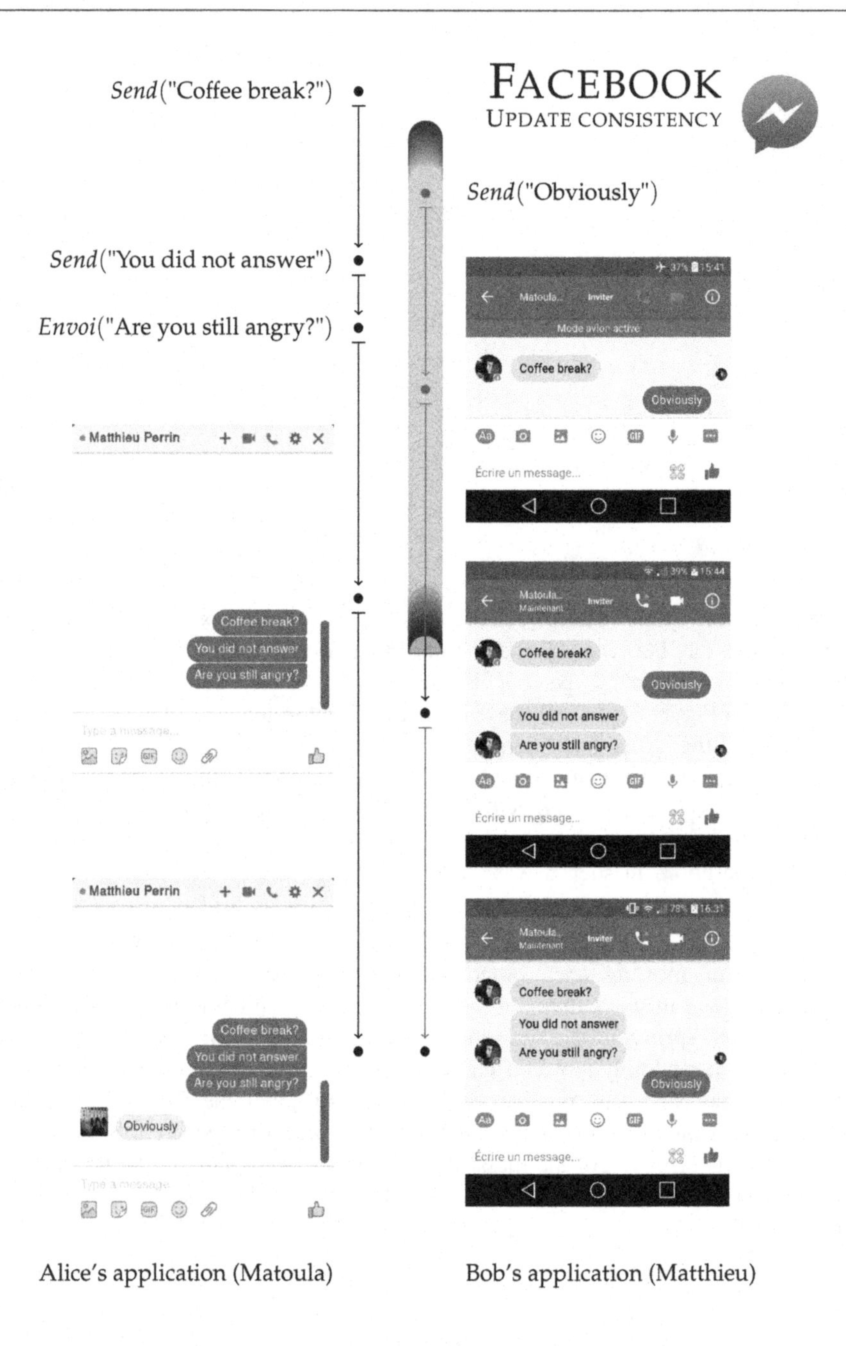

Figure 3.1. *Facebook Messengers behavior after a disconnection*

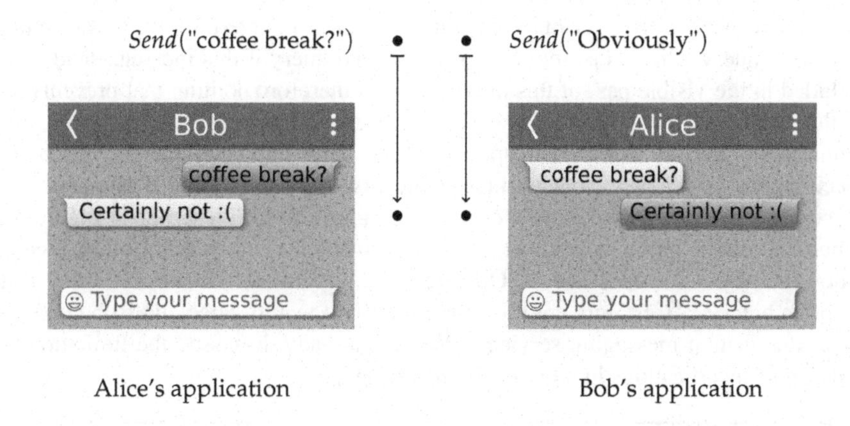

Figure 3.2. *A history not verifying update consistency*

3.2. Introduction

The observed histories generated by Skype (p. xv), Facebook Messenger, and the April fool prank are all eventually consistent (see p. 37): during their last query, Bob and Alice's applications have properly returned the same messages in the same order. Yet, these three applications offer very different user experiences.

The question that arises is naturally the following one.

Problem. *How can eventually consistent shared objects be specified?*

In the history of the conversation with Facebook Messenger in Figure 3.1, eventual consistency is achieved after a very long delay. Strong eventual consistency (see p. 39) is designed to prevent this situation: during her second and third queries, Alice received the same messages from Bob, she should therefore read the same state. The contribution of strong eventual consistency compared with eventual consistency is actually too small to be that significant. It turns out that the number of possible visible pasts exponentially increases with the number of updates in the history. In eventually consistent histories, as the number of updates increases, it becomes easier to find a different visible past for all queries performed before eventual consistency. In this history, we can imagine that the visible past of Bob's second query contains only Alice's first two sends and his own, while the visible past of his third query contains all the sends. Since Bob's last two queries do not have the same visible past, it is acceptable that they return different results.

The bias used in the previous reasoning is that it is counterintuitive that Bob may be able to query Alice's last message in his second query if this message send is not included in the visible past of this query; there is therefore nothing that prevents this in the definition. This absense makes it necessary to use additional specification techniques, such as concurrent specifications. This approach merely defers the specification problem to concurrent specifications: for example, it is possible to propose a concurrent specification of the messaging service presented in Figure 3.2, which indicates that if a process has knowledge of a message "Coffee break?" followed by a message "Obviously", it must be in the state ["Coffee break?"; "Certainly not:("]. Therefore, why would Alice and Bob complain about this instant messaging service? We have already discussed the limitations of concurrent specifications in more detail in section 2.4.5.

Approach. *We propose to give sequential specifications their place back by adding a constraint to eventual consistency: the convergence state must be compatible with a sequential execution of all updates in the history. We call this new consistency criterion* update consistency *and introduce* strong update consistency *as the equivalent for update consistency, of strong eventual consistency for eventual consistency.*

These new criteria are illustrated with the histories given in Figure 3.3, in which two processes share an instance of the integer set ($\mathcal{S}_{\mathbb{N}}$, see p. 5). In all the histories, the first process inserts the value 1 (operation I(1)), deletes the value 2 (operation D(2)) then reads an infinite number of times (operation R). The second process inserts the value 2, deletes the value 1 and reads an infinite number of times.

Their implementation is an important issue for justifying the introduction of these new criteria. We give special attention to this issue in this chapter by studying three generic algorithms offering different complexities. The first is effective at taking into consideration the number of exchanged messages, but requires an unbounded quantity of memory and computing power. The second solves this problem by increasing the number of exchanged messages. The third is parameterized to achieve a compromise so as to make the best usage of memory and network capacity.

This chapter is organized as it follows. Section 3.3 presents the two new criteria: update consistency [PER 14] and strong update consistency [PER 15], and compares them with eventual consistency and strong eventual consistency. Then, section 3.4 presents three algorithms that can be adapted to implement all abstract data types.

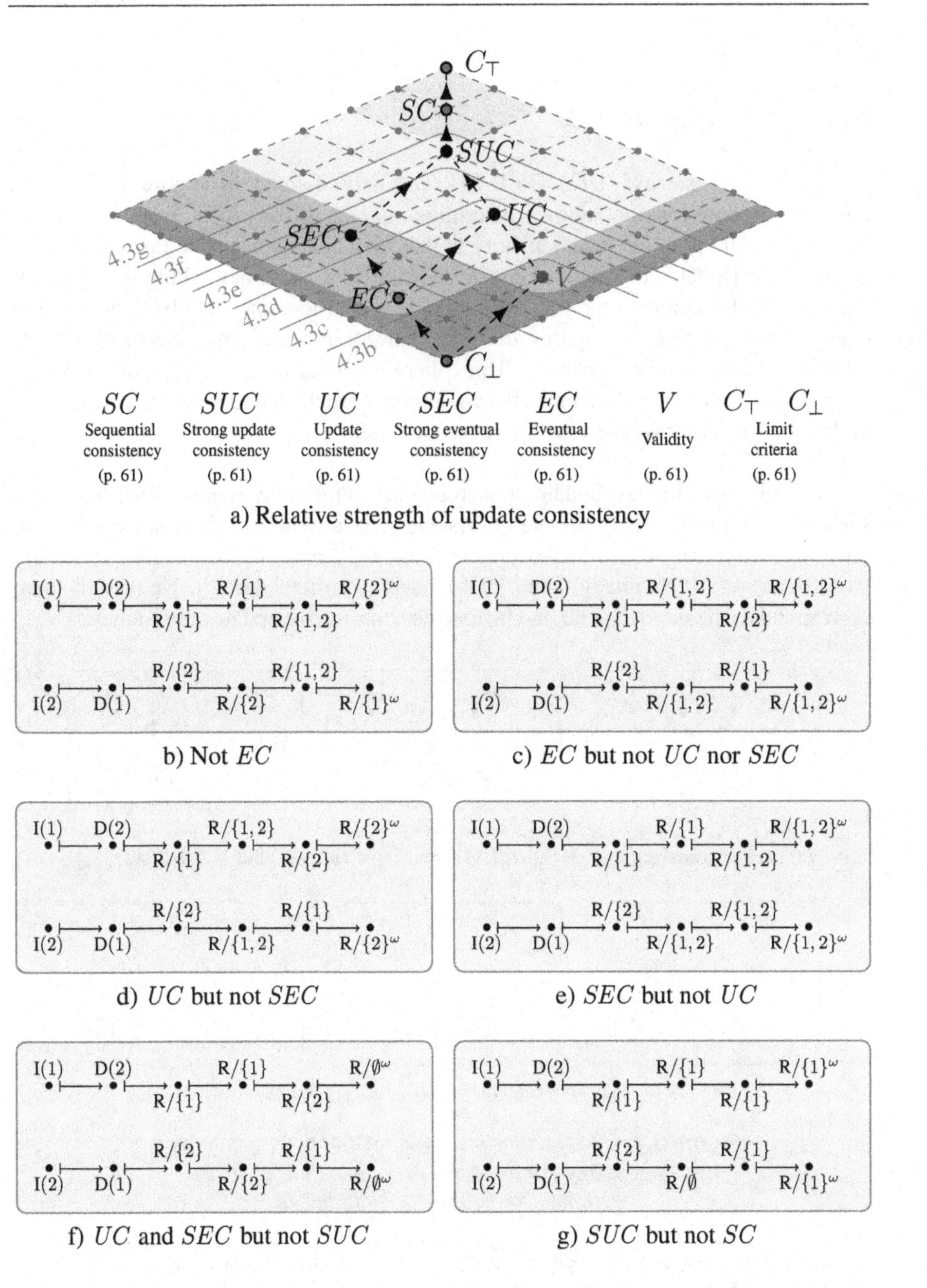

Figure 3.3. *Examples of histories for update consistency. For the color version of this figure, see www.iste.co.uk/perrin/distributed.zip*

3.3. Consistency criteria

3.3.1. *Update consistency*

Update consistency (UC) [PER 14], formally defined in Figure 3.5, is the consistency criterion that allows discriminating the history of Carole's prank in Figure 3.2. The problem in this history is that the convergence state does not reflect the updates. Update consistency reinforces eventual consistency taking into account the sequential specification of the abstract data type. More specifically, in addition to eventual consistency, it requires that the convergence state be one of the states allowed by sequential consistency. Thus, there is a total order on all updates (hence the name of the criterion) that satisfies the process order such that the execution of all updates in this order leads to the convergence state.

Assuming that the last update of each process is infinitely repeated, the history in Figure 3.2 does not verify update consistency. As a matter of fact, since there are only two ways for ordering two updates, the only possible convergent states are ["coffee break?"; "Obviously"] and ["Obviously"; "coffee break?"]. Neither of them correspond to the queried state, the history does not verify update consistency.

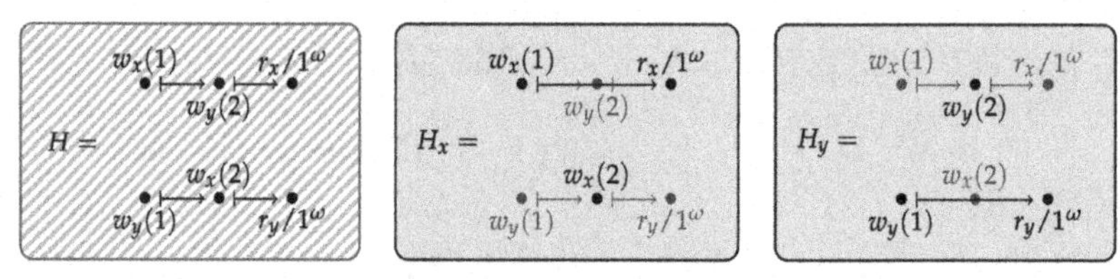

(a) Lack of composability: $H_x \in UC(\mathcal{M}_x)$ and $H_y \in UC(\mathcal{M}_y)$ but $H \notin UC(\mathcal{M}_{\{x,y\}})$

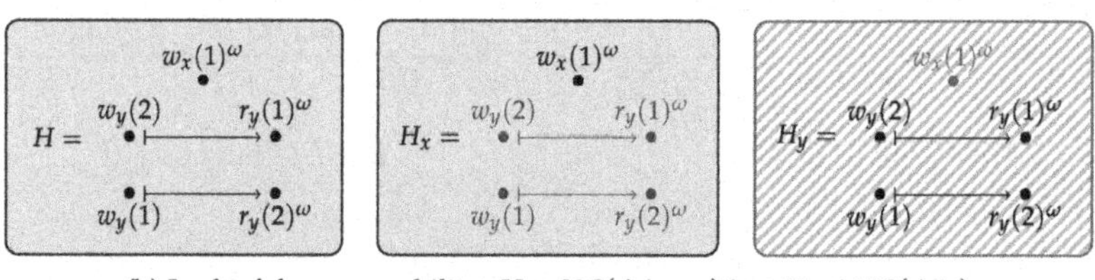

(b) Lack of decomposability: $H \in UC(\mathcal{M}_{\{x,y\}})$ but $H_y \notin UC(\mathcal{M}_y)$

Figure 3.4. *Update consistency is neither composable nor decomposable. For the color version of this figure, see www.iste.co.uk/perrin/distributed.zip*

Update consistency is neither composable nor decomposable, as Figure 3.4 illustrates it. In Figure 3.4(a), two processes share two registers named x and y. The scenario is very similar to the histories given in Figure 3.3: each process writes 1 into one of the registers and then 2 into the other. Convergence towards the state in which both registers are equal to 1 is prohibited due to update consistency. Yet considering each register separately, the update of the 2 could be placed before that of the 1. Thus, the two sub-histories H_x and H_y verify update consistency, but not history H. Figure 3.4(b) shows that update consistency is not decomposable using the same technique as for eventual consistency (p. 38): the proposed history comprises an infinite number of writes to x, which is sufficient to ensure update consistency regardless of reads performed to y.

PROPOSITION 3.1.– *Update consistency reinforces eventual consistency.*

PROOF.– Let an abstract data type be $T = (\mathsf{A}, \mathsf{B}, \mathsf{Z}, \zeta_0, \tau, \delta)$ and a concurrent history $H = (\Sigma, E, \Lambda, \mapsto)$ that verifies update consistency with regard to T (i.e. $H \in UC(T)$). We show that H is eventually consistent (i.e. $H \in EC(T)$).

If H contains an infinite number of updates or a finite number of pure queries, it is eventually consistent by definition. We assume that H contains a finite number of updates and an infinite number of pure queries. Since H verifies update consistency, there is a cofinite set of events $E' \subset E_H$ and a word $l \in \text{lin}(H[E_H/E']) \cap L(T)$. Since the number of updates is finite, there is a finite prefix l' of l that contains them all. Since $l' \in L(T)$, it labels a path between ζ_0 and a certain state ζ in the ADT. All events present in l but not in l' are pure queries, therefore they are all carried out in the state ζ from which $H \in EC(T)$. □

3.3.2. *Strong update consistency*

Whereas the history of Carole's prank (Figure 3.2) contains a finite number of queries, in contrast, it verifies both strong eventual consistency and update consistency, but for different reasons: strong eventual consistency because the sequential specification is not taken into account and update consistency because the convergence instant is not yet reached. Strong update consistency (SUC) [PER 15], formally defined in Figure 3.6, is to update consistency what strong eventual consistency is to eventual consistency. Similarly to strong eventual consistency, strong update consistency requires that two operations with the same past according to the visibility relation be performed in the same state, and just as with update consistency that state must be the result of a linearization of the updates of their common past. In the history given in Figure 3.2, regardless of its visible past in both message transmissions, a query will never be able to return the "Certainly not:(" message.

PROPOSITION 3.2.– *Strong update consistency strengthens update consistency.*

PROOF.– Let an abstract data type be $T = (\mathsf{A}, \mathsf{B}, \mathsf{Z}, \zeta_0, \tau, \delta)$ and a concurrent history $H = (\Sigma, E, \Lambda, \mapsto)$ that verifies strong update consistency with regard to T (i.e. $H \in SUC(T)$). We show that H verifies update consistency (i.e. $H \in UC(T)$).

UPDATE CONSISTENCY

✗ Not composable p. 42
✗ Not decomposable p. 42
✓ Weak p. 142

Similarly to eventual consistency, update consistency [PER 14] requires that if all participants stop updating, all replicas end up by converging to a common state. In addition, the possible convergence states are exactly those that can be achieved with sequential consistency: they must be the result of a linearization of the updates. For example, in the history of Figure 3.3d (recalled opposite), after some time, all the queries return the value $\{2\}$ *which can be obtained with the sequence* $I(1) \cdot D(2) \cdot I(2) \cdot D(1)$.

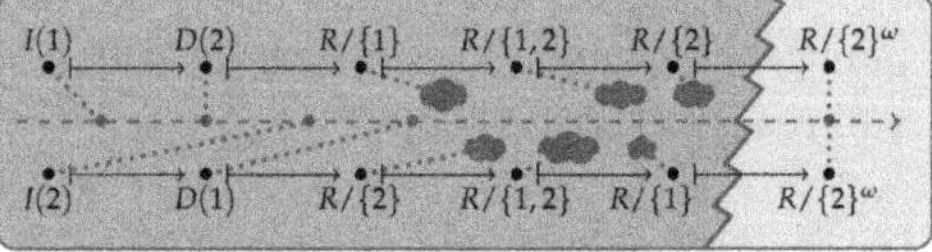

Formally, a history H satisfies update consistency for an abstract data type T (Definition 3.1) if it falls within one of the following two cases.

– *H contains an infinite number of updates (i.e.* $|U_{T,H}| = \infty$*). In this case, at least one participant never stops querying, therefore the values returned by the queries can continue to evolve according to the updates.*

– *It is possible to remove a finite number of queries (those made before convergence) to obtain a sequentially consistent history. In Definition 3.1, these queries are contained in the set* $E_H \setminus E'$ *(which must be finite) and the history* $H[E'/E_H]$ *contains all updates (in* E_H*) and the queries made after convergence (which belong to* E'*).*

In the example above, one of the possible values for E' *contains only the queries returning* $\{2\}$ *at the end, and the following sequence is a possible linearization of* $H[E'/E_H]$*. It should be noted that the input symbols of queries R are indeed present in the linearization because the operations are only hidden, but they play no role in the sequential specification since they correspond to loops on the states.*

$$I(1) \cdot D(2) \cdot I(2) \cdot D(1) \cdot R \cdot R \cdot R \cdot R \cdot R \cdot R \cdot R/\{2\}^{\omega}$$

DEFINITION 3.1.– Update consistency *is the consistency criterion*

$$UC : \left\{ \begin{array}{ccl} \mathcal{T} & \rightarrow & \mathcal{P}(\mathcal{H}) \\ T & \mapsto & \left\{ H \in \mathcal{H} : \begin{array}{l} |U_{T,H}| = \infty \\ \vee \ \exists E' \subset E_H, \ |E_H \setminus E'| < \infty \\ \quad \wedge \ \mathrm{lin}(H[E'/E_H]) \cap L(T) \neq \emptyset \end{array} \right\} \end{array} \right.$$

Update consistency guarantees two properties:

Convergence is obtained. *The history of Figure 3.3b does not satisfy update consistency because an infinite number of queries return 1 and another infinite number of queries return 2, although the history only contains six updates.*

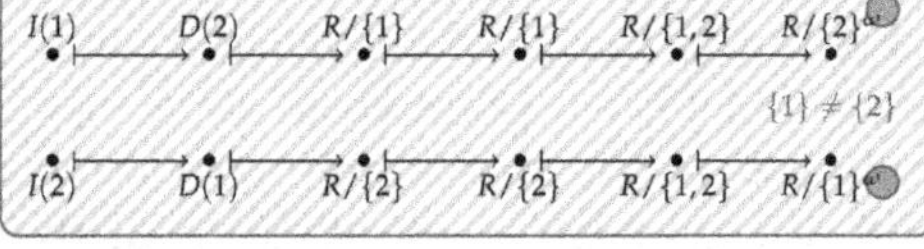

The process order between the updates is satisfied. *The history of Figure 3.3c does not satisfy update consistency. Indeed, all the queries achieved after convergence are convergent in the state* $\{1, 2\}$*. Moreover, in any linearization, the last update is a delete. The only convergence states admissible for this history are* $\emptyset, \{1\}$ *and* $\{2\}$.

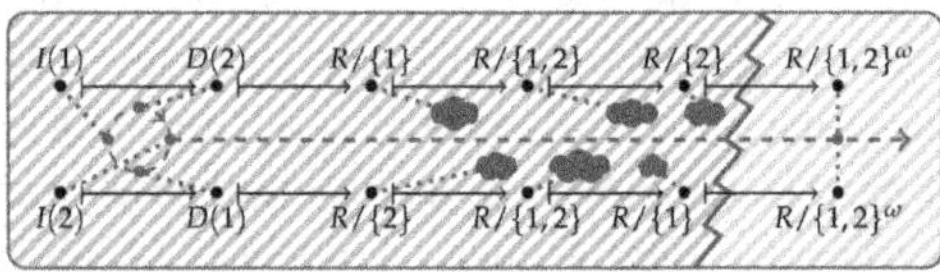

Figure 3.5. *Update consistency*

STRONG UPDATE CONSISTENCY

✗ Not composable p. 42
✗ Not decomposable p. 42
✓ Weak p. 142

Strong update consistency [PER 15] is to update consistency what strong eventual consistency is to eventual consistency. Just like strong eventual consistency, strong update consistency requires that two operations with the same visibility be made in the same state. Similarly to update consistency, update operations are totally ordered. The queried state for any event is the result of executing its past updates according to the visibility relation, ordered according to the total order. In the history of Figure 3.3g recalled here, the visibility relation is depicted by dotted lines and the total order by the order of the four points that precede each operation. The queries made before convergence are the result of a subsequence of the update sequence, the "holes" (represented by white circles) being filled as the execution proceed.

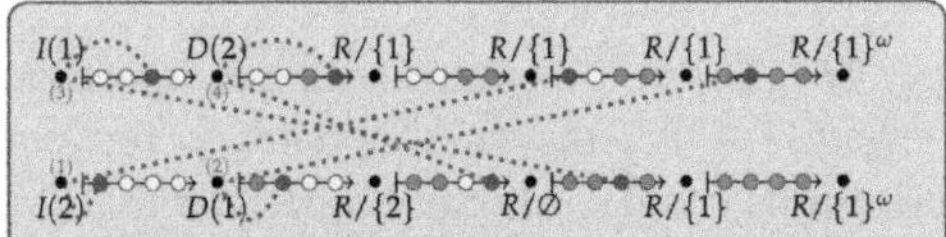

Formally, a history H verifies strong update consistency for an abstract data type T (Definition 3.1) if there is a visibility relation $\xrightarrow{VIS}$ (which models message reception, see page 67) and a total order $\leq$ on the events of H that contain $\xrightarrow{VIS}$ such as the history $\left(H^{\leq}[\lfloor e \rfloor_{\xrightarrow{VIS}} / \{e\}]\right)$, including the past updates of e according to $\xrightarrow{VIS}$ ordered following $\leq$ and e as unique potential query is sequentially consistent. In the example above, the total order is such that $I(2) \leq D(1) \leq I(1) \leq D(2)$ (mixing events and their labels) and the following sequences are the linearizations for the second process (ignoring the queries):

$I(2)\cdot D(1)\cdot R/\{2\}$ $\quad$ $I(2)\cdot D(1)\cdot D(2)\cdot R/\emptyset$ $\quad$ $I(2)\cdot D(1)\cdot I(1)\cdot D(2)\cdot R/\{1\}$

DEFINITION 3.2.– Strong update consistency *is the consistency criterion*

$$SUC : \begin{cases} T \to & \mathcal{P}(\mathcal{H}) \\ T \mapsto & \left\{ H \in \mathcal{H} : \begin{array}{l} \exists \xrightarrow{VIS} \in Vis(H), \exists \leq \text{ total ordering on } E_H, \xrightarrow{VIS} \subset \leq \\ \wedge \quad \forall e \in E_H, \text{lin}\left(H^{\leq}[\lfloor e \rfloor_{\xrightarrow{VIS}} / \{e\}]\right) \cap L(T) \neq \emptyset \end{array} \right\} \end{cases}$$

Strong update consistency is strictly stronger than UC + SEC. In effect, the history of Figure 3.3f (recalled opposite) verifies update consistency and strong eventual consistency but not strong update consistency. The problem occurs during the query of the 1 by the second process. For the convergence state to be $\emptyset$, it is necessary that $I(1) \leq D(1)$. However, the delete of the 1 precedes the query to the 1 according to the visibility relation and it is impossible that there is another insertion of 1 between the delete and the query.

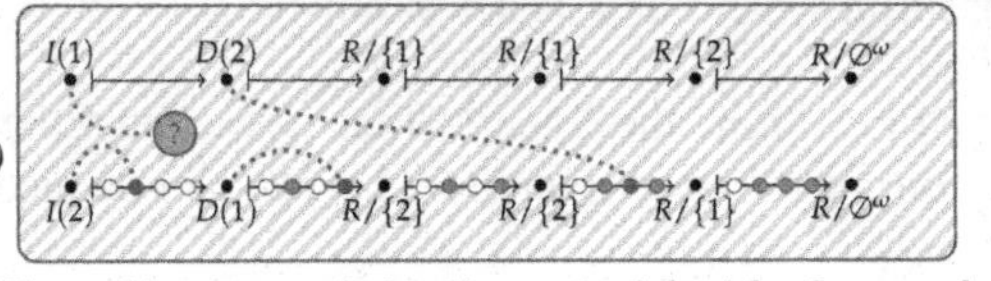

Figure 3.6. *Strong update consistency*

If H contains an infinite number of updates or a finite number of pure queries, it verifies update consistency by definition. We assume that H contains a finite number of updates and an infinite number of pure queries, since H verifies strong update consistency, there is a visibility relation $\xrightarrow{VIS}$ and a total order $\leq$ that contains $\xrightarrow{VIS}$ such that for all $e \in E_H$, there is a linearization $l_e \cdot \Lambda(e) \in \text{lin}\left(H^{\leq}[\lfloor e \rfloor_{\xrightarrow{VIS}} / \{e\}]\right) \cap L(T)$.

Let E' be the set of events that contain all the updates in their past according to the total order: $E' = \bigcap_{u \in U_{T,H}} \{e \in E_H, u < e\}$. The complementary of E' is $E_H \setminus E' = \bigcup_{u \in U_{T,H}} \{e \in E_H, e \leq u\}$ that is a finite union (there are a finite number of updates in the history) of finite sets (according to the third condition in definition 2.4 and the fact that $\leq$ contains $\xrightarrow{\text{VIS}}$), and is thus finite.

History $H^{\leq}[E_H/E']$ contains a unique linearization l, since the order $\leq$ is total. For any event $e \in E'$, the same updates in the same order are present before e in the linearizations l and l_e. It can be derived that $l \in L(T)$ and therefore that $H \in UC(T)$. □

PROPOSITION 3.3.– *Strong update consistency strengthens strong eventual consistency.*

PROOF.– Let an abstract data type be $T = (\mathsf{A}, \mathsf{B}, \mathsf{Z}, \zeta_0, \tau, \delta)$ and a concurrent history $H = (\Sigma, E, \Lambda, \mapsto)$ that verifies strong update consistency with regard to T (i.e. $H \in SUC(T)$). We show that H is eventually consistent (i.e. $H \in EC(T)$).

Since H verifies strong update consistency, there is a visibility relation $\xrightarrow{\text{VIS}}$ and a total order $\leq$ that contains $\xrightarrow{\text{VIS}}$ such that for all $e \in E_H$, there is a linearization $l_e \cdot \Lambda(e) \in \text{lin}\left(H^{\leq}[\lfloor e \rfloor_{\xrightarrow{\text{VIS}}} / \{e\}]\right) \cap L(T)$.

For all $e \in E_H$, we denote $V_e = \left(U_{T,H} \cap \lfloor e \rfloor_{\xrightarrow{\text{VIS}}} \setminus \{e\}\right)$. We define function f that associates with any set of updates E_H $V \subset \mathcal{P}(U_{T,H})$ the state obtained by executing all the updates of V in the order $\leq$. This function is well defined because the transition system of T is deterministic.

Given $e \in E_H$, the linearization l_e executes the updates of V_e in the order $\leq$. It therefore results in the state $f(V_e)$. We can deduce that $f(V_e) \in \delta_T^{-1}(\Lambda(e))$, thus $H \in SEC(T)$. □

3.3.3. *Case study: the shared set*

The set is probably the most studied data structure regarding eventual consistency. Various extensions of CRDTs have been proposed to implement the eventually consistent set although the insertion and deletion operations of the same element are not commutative. Among them, the OR-Set is an implementation of the shared set specified by the IW-Set concurrent specification (see p. 41). The purpose of this section is to compare the OR-Set and the strong update consistent set S_{Val} (see p. 29).

In the OR-Set, when an insertion and a deletion of the same element are concurrent, the deletion is canceled, which corresponds to the "insertion wins" strategy. Two properties restrict the set of histories admitted by the OR-Set: its consistency criterion SEC (all histories that it admits are in $SEC(\mathcal{S}_{Val})$) and its IW-Set concurrent specification (some histories of $SEC(\mathcal{S}_{Val})$ are not allowed). Our proof that strong eventual consistency is weaker than strong update consistency does not apply to the OR-Set, because the specification of the latter is reinforced by the concurrent specification. We now prove that this remains true for the OR-Set.

PROPOSITION 3.4.– *Let H be a history verifying strong update consistency for the set $\mathcal{S}_{Val}$. Then H is strongly eventually consistent for the IW-Set.*

PROOF.– Let $H \in SUC(\mathcal{S}_{Val})$. We define the new relation $\xrightarrow{IW}$ as follows. For all $e, e' \in E_H$, $e \xrightarrow{IW} e'$ if one of the following conditions holds:

– $e \xrightarrow{\text{VIS}} e'$;

– e and e' are two updates on the same element, i.e. $e \leq e'$ and there exists $x \in Val$ such that $\{\Lambda(e), \Lambda(e')\} \subset \{I(x)/\bot, D(x)/\bot\}$;

– e' is a query and there is an update e" such that $e \xrightarrow{IW} e''$ and $e'' \xrightarrow{IW} e'$.

The relation $\xrightarrow{IW}$ is acyclic because it is contained in $\leq$ and it is indeed a visibility relation since $\xrightarrow{\text{VIS}} \subset \xrightarrow{IW}$. In addition, it is not possible that two updates concerning the same element are concurrent according to $\xrightarrow{IW}$ and the last update for each element is the same as according to $\leq$. Therefore, H satisfies the strong eventual consistency property for the IW-Set. □

The natural question that then arises is that concerning the specification of the OR-Set in our formalism. The following three approaches are possible.

OR consistency. The first possible approach is to create a new consistency criterion, say OR, such that $OR(\mathcal{S}_{Val})$ describes very precisely the OR-Set. It should be noted that whereas such an approach is still possible in theory, it is difficult in practice to give meaning to such a new criterion, since it is necessary to identify the way in which concurrency is managed without reference to the particular data type.

Cache consistency. The second possibility is to sub-specify the OR-Set by using a consistency criterion slightly weaker than the one that the implementation provides in practice. This is what is achieved by bringing forward the eventual consistency guaranteed by the OR-Set. Furthermore, cache consistency $SC^\star$ more precisely describes the OR-Set, since it is always possible to order the insertion, deletion and query operations relating to a particular element of the set. However, the OR-Set guarantees more than cache consistency in practice.

Update consistency. The history in Figure 3.3(e) is the typical example of behavior admitted by the OR-Set but not by update consistency. Nonetheless, the justification often provided by the OR-Set is that it seems curious to be willing to delete an element that does not exist in the set, and thus that such a situation is only theoretical and not disturbing in practice. In fact, this is the case for many applications. For example, in a shared file system, the deletion of a file can only take place after its creation. The same is true concerning the deletion of a character in a collaborative editor. The third way to specify the OR-Set is thus to describe the kind of application being targeted: the OR-Set is an approximation of the set verifying update consistency, correct under certain assumptions which still have to be clarified.

3.4. Generic implementations

The purpose of this section is to study generic constructions that implement update consistency for any ADT in wait-free systems. In addition to the evidence that this criterion is indeed a weak criterion, this construction provides general techniques which can be combined to implement update consistency.

Two strategies are possible. We begin by presenting them both (UC_∞ in section 3.4.1 and UC_0 in section 3.4.2), before focusing on the parameterized algorithm $UC[k]$ that combines the two methods in order to extract the best aspects (section 3.4.3). The integer parameter k allows us to focus on one or the other strategy: $UC[0]$ is an optimization of UC_0 and the behavior of $UC[k]$ approximates that of UC_∞ when k tends to infinity.

3.4.1. *The UC_∞ algorithm for strong update consistency*

The algorithm in Figure 3.7 presents an implementation of the first strategy. An example of its execution is shown in Figure 3.8. The updates are denoted by the letters a, b, c and d and the states are named according to the sequence of operations that leads thereto (e.g. a shifts from the state ζ_0 to the state ζ_a, b shifts from the state ζ_a to the state ζ_{ab}, etc.).

The algorithm can be compared with Karsenty and Beaudouin-Lafon's [KAR 93] for objects including an *undo* operation canceling the last operation. The principle is to build a total order on the events accepted by all participants, and to rewrite the history *a posteriori* such that all processes reach the state corresponding to the same sequential history as soon as they have received all the updates of the concurrent history. Any of the strategies to create the total order would work. In Algorithm 3.7, this order is obtained from a Lamport linear logical clock [LAM 78] that contains the happened-before relation. The process order is thus necessarily respected. A Lamport clock is a total quasiorder since some events can be associated with the same clock. In order to have a total order, timestamps are pairs consisting of a linear clock and of the unique identifier of the process that has produced the event.

```
1  algorithm UC∞(A, B, Z, ζ0, τ, δ)
2      variable vtime_i ∈ ℕ ← 0;                          // linear logical clock
3      variable history_i ⊂ (ℕ × ℕ × A) ← ∅;                    // known updates
4      operation apply (α ∈ A) ∈ B
5          variable state ∈ Z ← ζ0;                                    // query
6          for (t_j, j, α′) ∈ history_i sorted according to (t_j, j) do
7              state ← τ(state, α′);                   // the history is replayed
8          end
9          if α ∈ U_T then
10             broadcast mUpdate (vtime_i + 1, i, α);                     // update
11         end
12         return δ(state, α);
13     end
14     on receive mUpdate (t_j ∈ ℕ, j ∈ ℕ, α ∈ A)
15         vtime_i ← max(vtime_i, t_j) ;
16         history_i ← history_i ∪ {(t_j, j, α)};
17     end
18 end
```

Figure 3.7. *Generic $UC_\infty(T)$ algorithm: code for p_i*

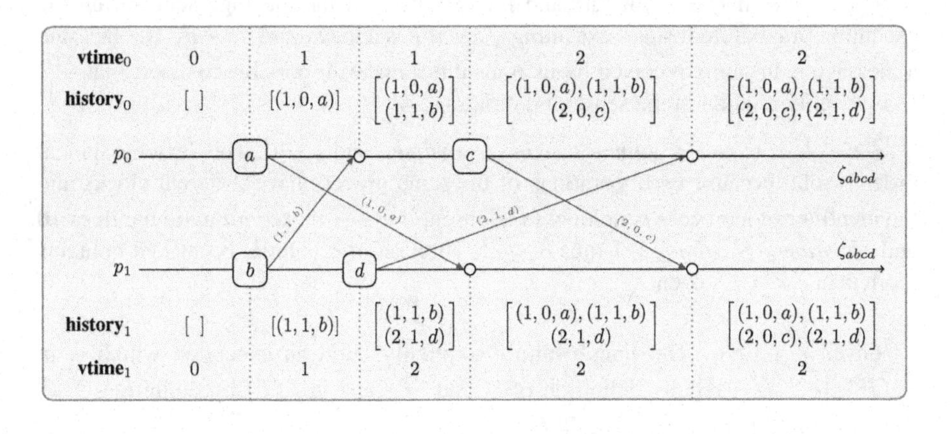

Figure 3.8. *Execution example of the algorithm UC_∞*

Each process p_i manages its own view $vtime_i$ of logical time and the list $history_i$ of known stamped update events by p_i. The list $history_i$ contains triplets (t_j, j, α) where α represents the update event and (t_j, j) is the timestamp. The list is sorted according to the lexicographic order of the timestamps: $(t_j, j) < (t_k, k)$ if $(t_j < t_k)$ or $(t_j = t_k$ and $j < k)$.

The algorithm executes the same code for all the operations. However, the `apply` method includes a query portion that can be ignored by pure updates and an update portion that is not necessary for pure queries. When an update is performed locally, p_i informs the other processes by reliably broadcasting a message to all processes (including itself). Therefore, all processes will eventually know all updates. When p_i receives a $\texttt{mUpdate}(t_j, j, \alpha)$ message, it updates its clock and inserts the triplet in its list $\mathit{history}_i$. To read the current state, p_i locally replays all updates contained in its list in the total order previously agreed upon, starting from the initial state.

A process never waits for messages from another process, which guarantees that all operations terminate even in the event of other processes crashing. The algorithm in Figure 3.7 correctly generates concurrent histories in all executions. We now show that these histories satisfy strong update consistency.

PROPOSITION 3.5.– *All the histories admitted by Algorithm 3.7 for an ADT T are in $SUC(T)$.*

PROOF.– Let $T \in \mathcal{T}$ and H be a history admitted by Algorithm 3.7. Given two events $e, e' \in E_H$ created by processes p_i and $p_{i'}$ in the respective states $(\mathit{vtime}_i, \mathit{history}_i)$ and $(\mathit{vtime}_{i'}, \mathit{history}_{i'})$. We define:

– $e \xrightarrow{\text{VIS}} e'$ if e is an update and $p_{i'}$ has received the message sent during the execution of e before it starts executing e', or if e is a query and $e \mapsto e'$. The fact that a message is instantly received by its transmitter makes it possible to assert that $\xrightarrow{\text{VIS}}$ is a visibility relation in the sense of definition 2.4.

– $e \leq e'$ if $\mathit{vtime}_i < \mathit{vtime}_{i'}$ or $\mathit{vtime}_i = \mathit{vtime}_{i'}$ and $i \leq i'$. This lexicographical order is total because two operations of the same process have different clocks and the identifier of a process is unique. In addition, if $e \xrightarrow{\text{VIS}} e'$, according to the lines 10 and 15, $\mathit{vtime}_{i'} \leq \mathit{vtime}_i + 1$ thus $e \leq e'$. The past of e is finite because it contains more than $c \times n + i$ events.

Given $e \in E_H$. The lines 6 and 7 explicitly build an execution which is in $\text{lin}(H^{\leq}[\lfloor e \rfloor_{\xrightarrow{\text{VIS}}} / \{e\}])$ by definition of $\leq$ and $\xrightarrow{\text{VIS}}$ and in $L(T)$ by definition of T. We thus correctly have $H \in SUC(T)$. □

3.4.2. UC_0 *algorithm*

The algorithm in Figure 3.9 presents the second solution. An example of its execution is given in Figure 3.10. The idea is that in pipelined consistency, when a process has received all the updates, its state is acceptable as a convergence state according to update consistency. Therefore, it is sufficient to choose a leader whose state is accepted by all other processes. Since it is impossible to know when all update operations have been carried out and what processes are correct, the leader's identity can vary over time. In the algorithm UC_0, the last leader is the correct process with the smallest identifier.

```
 1 algorithm UC0(A, B, Z, ζ0, τ, δ)
 2     variable clock_i ∈ ℕ[n] ← [0,...,0];                  // number of updates
 3     variable leader_i ∈ ℕ ← i;                             // leader's identity
 4     variable state_i ∈ Z ← ζ0;                               // current state
 5     operation apply (α ∈ A) ∈ B
 6         variable β ∈ B ← δ(state_i, α);                                // query
 7         if α ∈ U_T then
 8             FIFO broadcast mUpdate (clock_i[i] + 1, i, α);           // update
 9         end
10         return β;
11     end
12     on receive mUpdate (t_j ∈ ℕ, j ∈ ℕ, α ∈ A)
13         if clock_i[j] < t_j then                       // update not yet known
14             clock_i[j] ← t_j;
15             state_i ← τ(state_i, α);              // local application of α
16             leader_i ← i;
17             broadcast mCorrect (clock_i, i, state_i);
18         end
19     end
20     on receive mCorrect (cl_j ∈ ℕ[n], j ∈ ℕ, s_j ∈ S)
21         if ∀k, clock_i[k] ≤ cl_j[k] ∧ (j ≤ leader_i ∨ ∃k, clock_i[k] < cl_j[k]) then
22             clock_i ← cl_j;             // correction approval
23             leader_i ← j;
24             state_i ← s_j;
25         end
26     end
27 end
```

Figure 3.9. *Generic $UC_0(T)$ algorithm: code for p_i*

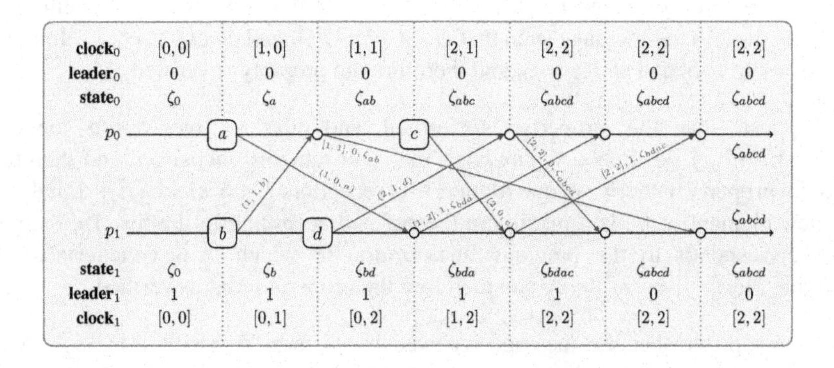

Figure 3.10. *Execution example of the UC_0 algorithm. For the color version of this figure, see www.iste.co.uk/perrin/distributed.zip*

Each process p_i manages three variables. The variable $clock_i$ contains a vector clock whose input $clock_i[j]$ represents the number of `mUpdate` messages sent by p_j and received by p_i. The local state upon which the queries are performed is $state_i$. Finally, if $leader_i \neq i$, $state_i$ contains a value that has been present in $state_{leader_i}$.

When executing an operation, Algorithm 3.9 reads its local state and broadcasts a `mUpdate`(cl_j, j, α) message if the operation is an update. This broadcast is FIFO to ensure that a process that never receives correction executes all operations in an order consistent with the process order. Here, it is important that the message is immediately received by p_i. When p_i receives the message from p_j, it verifies that the operation has not already been taken into account in its current state (in which case $clock_i[j] \geq cl_j$), and if this is not the case it applies the update to its local state, and then it assumes itself as its own leader and broadcasts a `mcorrect` message that contains the vector clock, its identifier and its new local state. When a correction is received, p_i can either choose to accept it or to refuse it. It accepts it if the correction does not cause it to lose information ($\forall k, clock_i[k] \leq cl_j[k]$) and if it allows it to acquire some more ($\exists k, clock_i[k] < cl_j[k]$) or if the process that sent the information is in a better position than itself to become leader ($j \leq leader_i$).

PROPOSITION 3.6.– *All the histories admitted by Algorithm 3.7 for an ADT T are in $UC(T)$.*

PROOF.– Let $T \in \mathcal{T}$ and H be a history admitted by the algorithm in Figure 3.9. If H involves an infinite number of updates or a finite number of queries, then $H \in UC(T)$. Otherwise, the number of updates performed by p_i is finite. We denote m_i, and let $cl_{max} = [m_0, ..., m_{n-1}]$. In addition, there is a time t_1 from which all correct processes have received all `mUpdate` messages and therefore a time t_2 from which they all received all the `mCorrect` messages.

From the sequence of states of p_i, we prove by induction that at any time, the state of p_i is the result of a linearization of the set U_i of the updates corresponding to a `mUpdate`(cl_j, j, α) message such that $cl_j \leq clock_i[j]$, and $clock_i \leq cl_{max}$. Initially, $clock_i = [0, ..., 0]$ and $state_i = \zeta_0$ and therefore the property is verified.

Suppose that the property is verified and that p_i receives a message `mUpdate`(cl_j, j, α). If $cl_j \leq clock_i[j]$, the state remains unchanged and therefore also the property. Otherwise, due to the FIFO reception, $cl_j = clock_i[j] + 1$ and thus the new element of U_i is a maximum thereof and according to the line 15, the new state corresponds to the previous linearization to which α is concatenated. In addition, $clock_i[j] = cl_j \leq m_j$ and therefore the property remains verified.

Now suppose that the message received by p_i is `mCorrect`(cl_j, j, s_j). After processing the message, either p_i is in the same state as before the reception or it is in the state of p_j at the time when it sent the message. The property is verified in every case.

In addition, at time t_2, for any j, $clock_i[j] = m_j$. It turns out that the clocks of the `mUpdate` messages sent by p_j are successive due to $clock_i[i] + 1$ in line 8, in line 14 and due to the fact that the messages are instantly received by their issuer. If the value c of $clock_i[j]$ at t_2 were $c < m_j$, p_j would thus have sent a `mUpdate`$(c + 1, j, \alpha)$ message, which would have been received by p_i before t_1 and according to line 14, $c \geq c + 1$, which is absurd.

Let p_i be a correct process. Based on the foregoing, at time t_2, its clock is equal to cl_{max}. At the time when its clock took the value cl_{max}, p_i had just received a message. If it is a `mUpdate` message, it has itself broadcast a `mCorrect` message. Otherwise, it has received a `mCorrect` message associated with the clock cl_{max}. In every case, a `mCorrect`(cl_{max}, k, ζ_k) has been broadcast. Among all these, consider the one for which k is minimal. When p_i receives this message, it has indeed $clock_i \leq cl_{max}$, since cl_{max} is the largest clock that p_i can reach during execution. Suppose that $clock_i = cl_{max}$ and $leader_i \leq k$. The previous modification of the state of p_i cannot be caused by the receiving of a correction because according to the definition of k, no `mCorrect`$(cl_{max}, leader_i, \zeta)$ messages have been broadcast. Neither can it be caused by the reception of an update because then p_i would have broadcast a message `mCorrect`$(cl_{max}, i = leader_i, \zeta)$, which is once again impossible. Thus, p_i has accepted the correction of p_k.

For any process p_i, the message `mCorrect`(cl_{max}, k, ζ_k) for which k is minimal is the same. We can deduce that after time t_2, for any process p_i, $clock_i = \zeta_k$. We also know that there is a linearization of all the updates that result in the state ζ_k. The set E' of the events performed before time t_2 is finite and $\text{lin}(H[E_H/E']) \cap L(T) \neq \emptyset$. The history H therefore verifies strong update consistency. □

3.4.3. *The* $UC[k]$ *algorithm*

The UC_∞ algorithm in Figure 3.7 is very effective concerning communication. A single message is broadcast for each update and every message contains only the information needed to identify the operation and a timestamp consisting of two integers; their size grows logarithmically only with the number of processes and the number of operations in the history. On the other hand, it is expensive in terms of memory and local computing time. As a matter of fact, the size of the local memory space and the time necessary to perform a query increase linearly with the number of operations.

Conversely, the UC_0 algorithm in Figure 3.9 is efficient in terms of computation time and memory. The state is immediately accessible for a query and the memory complexity is the same as for pipelined consistency, which is a state of the object and a vector of n integers in which the size of each one increases logarithmically with the number of operations. On the other hand, the algorithm is quadratic relatively to the number of messages broadcast per operation. That is, most sent messages include a vector clock and a local state. Nonetheless, the local state is likely to be very heavy. For example, in the case of a database, the local state is the totality of the database.

```
1  algorithm UC[k](A, B, Z, ζ0, τ, δ)
       // recent events
2      variable history_i ⊂ ℕ × ℕ × A ← ∅
       // Lamport's logic clock
3      variable vtime_i ∈ ℕ ← 0
       // summary of former events
4      variable rstate_i ∈ Z ← ζ0
       // limit between former and recent
5      variable rvtime_i ∈ ℕ ← 0
       // last former events
6      variable rclock_i ∈ ℕ[n] ← [0,...,0]
       // leader's identity
7      variable leader_i ∈ ℕ ← i
       // prevent multiple sends
8      variable sent_i ∈ 𝔹 ← false
9      operation apply (α ∈ A) ∈ B
10         variable q ∈ Z ← rstate_i;
           // Query portion management
11         if α ∈ Q_T then
12             for (t_j, j, α') ∈ history_i sorted according to (t_j, j) do q ← τ(q, α')
13         end
           // Update portion management
14         if α ∈ U_T then FIFO broadcast mUpdate (vtime_i + 1, i, α)
15         return δ(q, α);
16     end
17     function record (t ∈ ℕ)
18         rvtime_i ← max(rvtime_i, ((k > 0) ? (k × (⌊t/k⌋ − 1)) : (t)));
           // k ≤ t − rvtime_i ≤ 2k
19         for (t_j, j, α) ∈ history_i with t_j ≤ rvtime_i sorted according to (t_j, j) do
               // (t_j, j, α) becomes an former update
20             rclock_i[j] ← t_j; history_i ← history_i \ {(t_j, j, α)};
               rstate_i ← τ(rstate_i, α);
21             leader_i ← i; sent_i ← false;
22         end
23     end
24 end
```

Figure 3.11. *Generic $UC[k](T)$ algorithm: code for p_i (beginning)*

The $UC[k]$ algorithm in Figures 3.11 and 3.12 [RUA 14] combines the two previous methods by maintaining their best aspects. An example of its execution is shown in Figure 3.13. Real systems are rarely completely asynchronous. In general, asynchrony is used as an abstraction to model systems in which transmission times are actually bounded (at least with large probability) but the bound is too large to be used in practice (a user can be required to wait several minutes between each action in an interactive service) or unknown. This means that after a while, the prefix of the $history_i$ list in the UC_∞ algorithm will be fixed and it will be possible to summarize old messages in a state of the object. If an old message is received later than expected, the strategy of the UC_0 algorithm is used to guarantee update consistency. The total order imposed *a priori* by timestamps can be altered if all processes apply the same modification. This is done by exchanging the local states. Thus, the additional cost in memory and in computation time remains bounded depending on a parameter k that affects the size of the list. Furthermore, the additional bandwidth is only needed when messages are abnormally delayed, for instance after a partition. The limit of the $UC[k]$ algorithm when k tends to infinity is very similar to UC_∞ and the $UC[0]$ algorithm is an optimized version of UC_0, since the list always remains empty, but correction messages are only sent when updates occur.

```
1  algorithm UC[k](A, B, Z, ζ0, τ, δ)
2     on receive mUpdate (t_j ∈ N, j ∈ N, α ∈ A)
         // Lamport's clock management
3        vtime_i ← max(vtime_i, t_j)
         // update not yet known
4        if rclock_i[j] < t_j then
5           history_i ← history_i ∪ {(t_j, j, α)};
            // former update
6           variable conflict ∈ B ← t_j ≤ rvtime_i;
7           record(vtime_i);
8           if conflict then
9              broadcast mCorrect (rclock_i, rvtime_i, i, rstate_i);
               sent_i ← true;
10          end
11       end
12    end
13    on receive mCorrect (cl_j ∈ N[n], t_j ∈ N, j ∈ N, q ∈ Z)
14       if rvtime_i < t_j then record(t_j + k)
15       if ((rclock_i < cl_j) ∨ (rclock_i = cl_j ∧ j < leader_i)) then
            // accept the correction
16          rclock_i ← cl_j; rstate_i ← q; leader_i ← j; sent_i ← true;
17       else if
          (cl_j < rclock_i ∨ (cl_j = rclock_i ∧ leader_i < j)) ∧ j ≠ i ∧ ¬sent_i
          then
            // refuse the correction and help p_j
18          broadcast mCorrect (rclock_i, rvtime_i, i, rstate_i); sent_i ← true;
19       end
20    end
21 end
```

Figure 3.12. *Generic $UC[k](T)$ algorithm: code for p_i (end)*

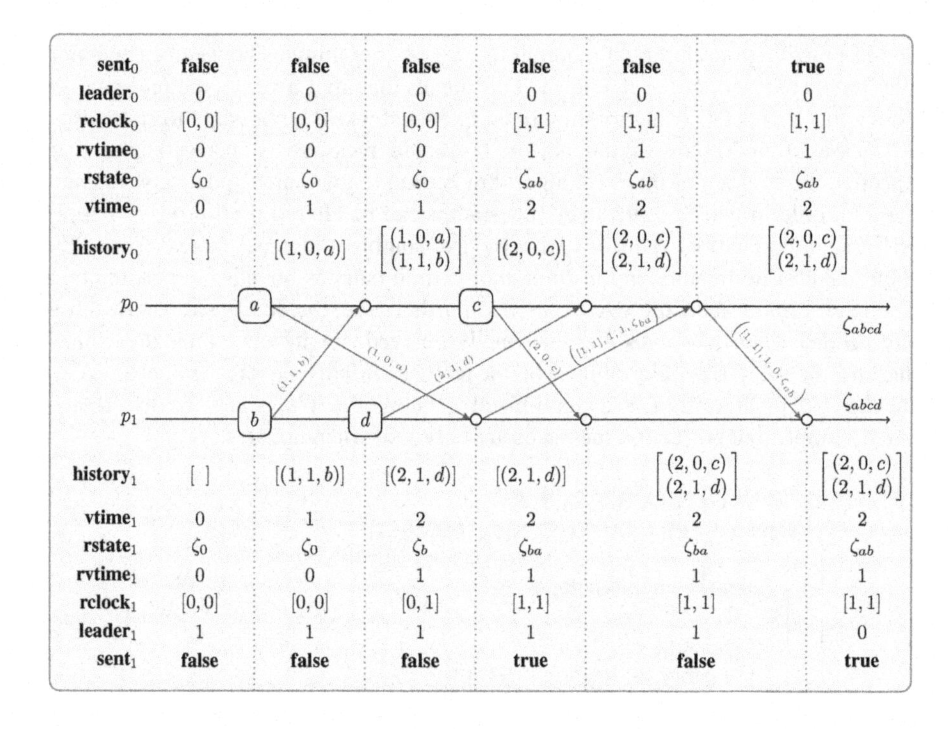

Figure 3.13. *Execution example of the $UC[k]$ algorithm. For the color version of this figure, see www.iste.co.uk/perrin/distributed.zip*

The process p_i manages the following variables:

– *history*$_i$ $\subset \mathbb{N} \times \mathbb{N} \times A \leftarrow \emptyset$: the list of "recent" updates, received but not yet implemented and of size $\mathcal{O}(n \times k)$;

– *vtime*$_i$ $\in \mathbb{N}$: the Lamport clock used to timestamp the updates;

– *rstate*$_i$ $\in Z$: the state that summarizes the set of the "former" updates that have been received but are no longer part of *history*$_i$;

– *rvtime*$_i$ $\in \mathbb{N}$: the limit date between recent and former updates. An update is said to be "former" if, and only if, the Lamport clock of its timestamp is smaller than *rvtime*$_i$. Otherwise, it is "recent";

– *rclock*$_i$ $\in \mathbb{N}[n]$: the vector clock associated with *rstate*$_i$, used to compare states. The number *rclock*$_i[j]$ represents the largest linear clock of the updates achieved by p_j and applied to build *rstate*$_i$;

– *leader*$_i$ $\in \mathbb{N}$: the identifier of the last process of which p_i has accepted a correction;

– $sent_i \in \mathbb{B}$: a Boolean true if and only if the current state has already been broadcast, used to prevent multiple broadcasts of the same message.

The current state of a process is obtained by applying all the updates contained in $history_i$ at state $rstate_i$ in the lexicographic order of the timestamps.

To execute an update, p_i broadcasts in a FIFO manner a message consisting of the symbol of the operation and a timestamp (t_i, i) created similarly to the UC_∞ algorithm. When a process receives a $\texttt{mUpdate}(t_j, j, \alpha)$ message, it tries to insert it in its update list $history_i$. It then calls the method $record(t)$ which ensures that $vtime_i - rvtime_i < 2k$ and applies the updates whose timestamp is smaller than the new value of $rvtime_i$. A conflict is detected if the clock of the new message t_j is smaller than that of the state $rvtime_i$, which means that the update should already have been applied to $rstate_i$ — and has potentially been by another process. To solve the conflict, p_i locally applies the update to $rstate_i$ then informs the other processes of the conflict by sending its new state. As with the UC_0 algorithm, the linearization order remains correct due to the FIFO reception with `mUpdate` messages.

When p_i receives a $\texttt{mCorrect}(cl_j, t_j, j, q)$ message from p_j, it knows that p_j has observed a conflict and changed its linearization order. It can choose to accept or refuse the correction, as in the UC_0 algorithm, depending on the timestamps of its state and the correction, $rvtime_i$ and t_j:

– if $t_j < rvtime_i$, the state of the correction is very old. This case can occur if the correction message has been received with a lot of delay. If it accepted the correction, p_i would lose the information of the updates contained between t_j and $rvtime_i$ that it has already integrated into $rstate_i$. In turn, it broadcasts a correction message to help p_j catch up its delay;

– if $t_j > rvtime_i$, p_i is lagging compared with p_j. It then turns its own clock forward to t_j to fall within one of the following situations;

– if $t_j = rvtime_i$ and $cl_j = rclock_i$, the two states have been produced by taking into account the same updates but potentially in a different order. The arbitration is then performed by comparing the identifiers of the processors, and by using a variable $leader_i$ as in the UC_0 algorithm. If p_i decides to keep its own state, it broadcasts it so that other processors can make the same choice. Due to the variable $send_i$, a state can only be sent once;

– if $t_j = rvtime_i$ and $cl_j > rclock_i$, there are messages in transit destined to p_i, which will produce a new conflict in the future. To avoid future conflicts, p_i accepts the correction;

– if $t_j = rvtime_i$ and $cl_j < rclock_i$, similarly, p_j will generate a new conflict in the future, therefore p_i broadcasts its state to cancel out the errors of p_j;

– if $t_j = rvtime_i$ and both clocks are not comparable, choosing a state may cause information loss. In addition, we know that the two processes will have a new conflict in the future, therefore they will be able to agree later on. The correction is thus simply ignored.

In addition to the lack of waiting, the other aspect that guarantees tolerance to failures in this algorithm is that each process always keeps an acceptable state for a prefix of the history. The additional communication merely contributes to manage eventual consistency. Nevertheless, correct processes have all the time to agree (since they are correct), and faulty processes cannot perform an infinite number of updates, therefore they cannot prevent eventual consistency. The proposition 3.7 establishes that the $UC[k]$ algorithm is an implementation of update consistency. The complete proof of this proposition, as well as a study demonstrating the effectiveness of the algorithms, can be found in [PER 16a].

PROPOSITION 3.7.– *All the histories accepted by the algorithm in Figures 3.11 and 3.12 verify update consistency.*

3.5. Conclusion

This chapter introduces two new consistency criteria, update consistency and strong update consistency, which are respectively stronger than eventual consistency and strong eventual consistency and weaker than sequential consistency. In practice, an object verifying update consistency can be seen as a sequentially consistent object, whose state may be corrupt but able to repair itself after periods of instability. As the network gets faster and the frequency of updates on the object becomes smaller, the periods of instability become shorter. This criterion is therefore especially suited for systems controlled by humans capable of adapting to temporary errors that they can easily detect and correct. On the other hand, since any reading is likely to be erroneous, it is more difficult to program using objects verifying update consistency.

We have presented three generic algorithms to implement update consistency and strong update consistency in wait-free systems. The first, UC_∞ is very communication efficient since it merely broadcasts a message for every update, but costly in terms of computing time and memory space because it maintains the entire history in memory and replays it for each query. On the contrary, UC_0 is expensive communication-wise but inexpensive with space and computing time. The third offers a compromise between the first two to be efficient with regard to both space and communication. It is parameterized by an integer k that determines the maximal space required in memory. For $k = 0$, this is an optimization of UC_0 and the larger k becomes, the closer to UC_∞ its behavior becomes. There is an optimal value of k depending on the ratio of the average message transmission speed on the network and the frequency of update operations, for which the algorithm does not make use of more bandwidth than UC_∞ while using minimum memory space. The problem is that the calibration of k is difficult *a priori*. A research avenue would be to allow the algorithm to modify k on the fly to optimize itself during execution. The difficulty lies in maintaining an identical k in all processes.

Another research avenue concerns object-specific implementations in particular. The particular structure of certain abstract data types allows optimizations that make a specific implementation more effective than the generic implementation. This is true, for example, in the case of the set. In the UC_0 algorithm, it is important to maintain the complete history in memory because the state of an object with any type may potentially depend on all the operations present in the history. For the set, only the last insertion or deletion operation on each element has an impact on the read state. Therefore, only the set of the elements inserted at least once can be kept in memory associated with their last update timestamp.

4

Causal Consistency

4.1. Preamble

In order to grasp why Bob was so upset with Alice, we need to understand his special relationship with his cat. The absence of any welcoming miaows when he got home one evening gave him a shock which aggravated his natural anxiety. Later that evening, the absence of the familiar purring and weight on his knees encouraged him to organize a search mission to find his four-legged acolyte. This important mission was organized over instant messaging with his two best friends, Alice and Carole. In order to appease Carole, he used the application she had developed which, once the anticipated surprise died away, had been modified to transmit messages correctly. However, the call for help did not have the expected effect. Alice responded point-blank that he was lucky his cat was gone! In the face of this cynicism, he sent off a series of insults and quit the session. Alice could not understand Bob's reaction. All she had done, after all, was respond to Carole's message telling her that Carole was in New York for a week on vacation, which was reason enough to rejoice. Alice was completely ignorant of her childhood friend's misfortunes.

The two young women continued to talk for some time and realized that Bob hadn't received Carole's message. One question remained unanswered: if Bob had stayed on a little longer, he would surely have received Carole's message? This might have pacified him...if his messages had been reordered as Skype reorders messages. Bob would then have realized that Alice's message was in response to Carole's. However, if the contrary had happened he would have thought that Carole was mocking him.

4.2. Introduction

The histories given in Figure 4.1 summarize the stages of the conversation (these images are not taken from real-life incidents). Bob had initially sent Alice and Carole the message, “Any volunteers to help find my cat? :’(”. Carole had simultaneously sent her message “I’m in New York all week :-/”. There was a lag between the message being sent by Bob and Alice receiving the message and, consequently, Alice received Carole’s message before Bob’s message. From her point of view, therefore, there was no ambiguity about what her response meant when she said: “Aren’t you lucky XD”. However, Bob received Alice’s message before Carole’s message and he thought that Alice had responded directly to him.

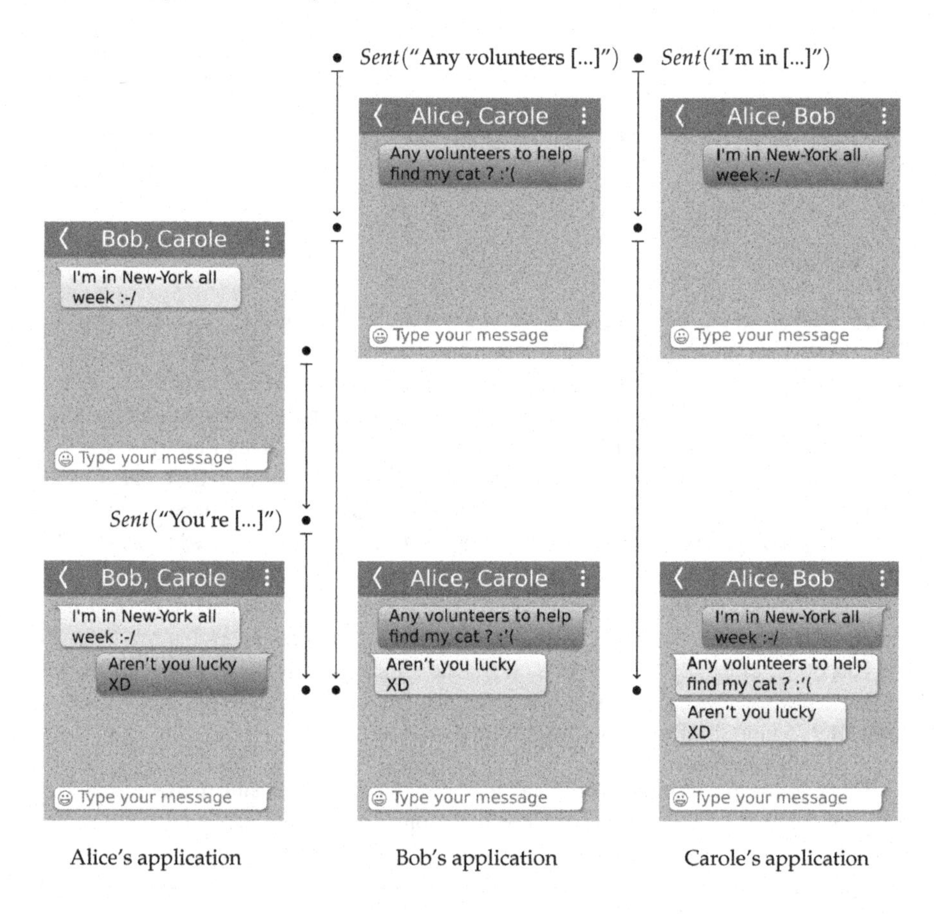

Figure 4.1. *Is Alice talking about Bob's cat or Carole's vacation?*

The problem in this story is temporal order: Alice responded to Carole’s remark after having read it and she couldn’t have read it until Carole sent it. Alice’s message was sent after the message sent by Carole and it is strange that Bob received her message before Carole’s. The order intuitively presented here is generally called *causal order*.

Causal memory (see p. 50) is a stronger property than PRAM (the application of pipelined consistency to memory), which prevents similar behavior for memory. To know whether the history verifies this property, we must try to construct a partial order, called *causal order*, which contains the process order, and a *writes-into* order, which encodes the fact that any value read from a register must have been previously written in the same register. The partial order thus created must be compatible with, for each process, a linearization containing the read of this process and the writes of all the processes.

Problem. *Two problems arise from the definition of causal memory.*

– Dependencies between the reads and writes are a central aspect in the definition of causal memory, but they are intimately linked to the sequential specification of the memory. How can you extrapolate the definition of causal consistency to other abstract data types?

– As causal order is only a partial order, concurrent writes are always possible. In concrete terms, the final discussion between Alice and Carole shows that the causality reinforces neither the pipelined consistency nor the update consistency and, therefore, we face the same problem that was mentioned in the general introduction. How can pipelined consistency and update consistency be adapted to take causality into account?

The natural approach used to extend causal consistency would be to define a concept of *semantic dependency* applicable to all abstract data types. The semantic dependencies of the memory would be dependencies between the reads and writes; for instant messaging services, the reading of a message would depend semantically on its sending, etc. In fact, this approach seems dangerous as the notion of semantic dependency is very vague given that each abstract data type has its own specificities. Thus, any attempt at a formal definition would be questionable. Let us look at two examples that illustrate the difficulties that may arise. It is, of course, possible to respond individually to each case by making the concept of semantic dependency more complex, but what about a general case?

The counter. Let us imagine the simplest counter, which only allows increments and reading. The value returned by a read does not depend specifically on any one write: it only indicates the number of writes that it depends on. How do we know which these are? The problem is also present for memory [MIS 86]: if the same value is written several times in the same register, how do you determine which write a read of that value depends on? We will see in section 4.4.2 that this question poses a real limitation to causal memory and the definition of this chapter will help you address this.

The queue. For the queue (see p. 9), each value returned by the *pop* operation depends semantically on a single call from the *push* operation, but all the operations carried out between these two events (and the order of these operations) are important in determining whether the value returned by the *pop* operation is valid. Semantic dependency thus happens through a chain of events.

Approach. *Rather than trying to find a general definition for semantic dependencies, our approach is the reverse: we require the existence of a partial order such that the events in the history verify these properties in relation to the object's sequential specification. A side effect of these properties is that the partial order must necessarily contain the semantic dependencies without us needing to define them.*

Small differences in the properties required of the partial order define different criteria. We will, therefore, study not only causal consistency, an extension of causal memory, but also the other variants of this criterion. This chapter defines four new consistency criteria:

– weak causal consistency may be seen as the lowest common denominator of the following three criteria. The conjunction of weak causal consistency and of convergence or pipelined consistency form new consistency criteria;

– causal convergence is obtained by combining weak causal consistency and eventual consistency. This is a stronger criterion than strong update consistency;

– causal consistency reinforces pipelined consistency. If we apply this to memory, it corresponds, under certain hypotheses, to causal memory;

– the standard implementation of causal consistency in message-passing systems implements slightly more than causal consistency. The behaviors allowed by the consistency criterion but which are not accepted by the implementations are often called "false causality". We will look at *strong causal consistency*, which reinforces causal consistency by taking into account any false causality situation.

Figure 4.2 illustrates the differences between them using a few examples in which three processes share an instance of sliding window register ($\mathcal{W}_2$, see p. 7). In all the histories (except that in Figure 4.2(i)), the first process writes 1 (operation w(1)) then reads an infinite number of times (operation r), the second process reads once then writes 2 and the third process writes 3 then reads an infinite number of times. Figure 4.2(a) summarizes all these criteria and their relative force.

The rest of this chapter is organized as follows: section 4.3 introduces the concept of *causal order*, central to all criteria related to causal consistency as well as weak causal consistency and causal convergence. Section 4.4 introduces causal consistency and strong causal consistency, as well as the link between causal consistency and causal memory. Finally, section 4.5 presents the application contexts in which these causal criteria behave like sequential consistency.

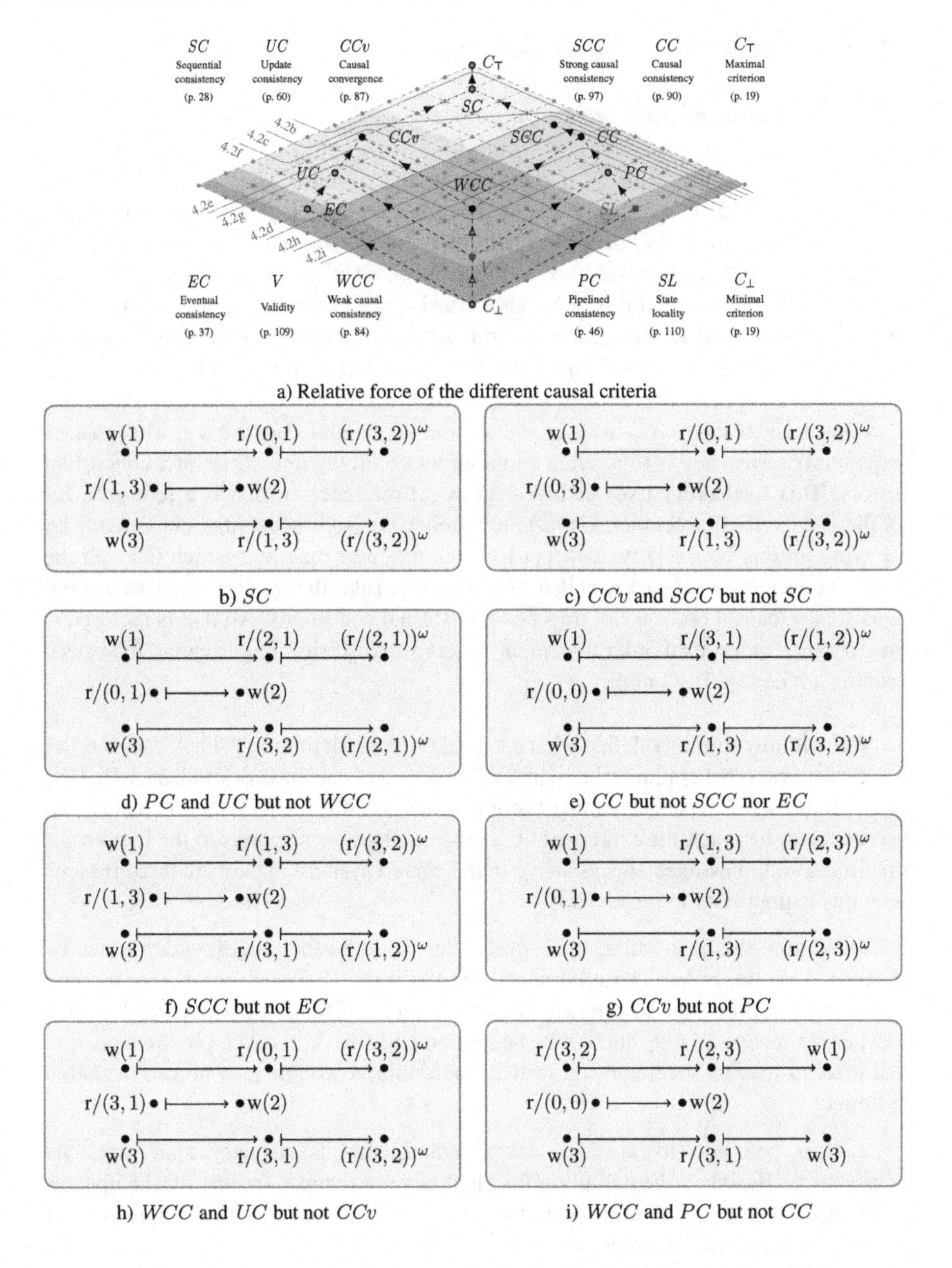

Figure 4.2. *Relative force of causal criteria. For the color version of this figure, see www.iste.co.uk/perrin/distributed.zip*

4.3. Causality as a consistency criterion

4.3.1. *Causal order and temporal cones*

Causality is often seen as a property of the system: causal broadcast (see section 1.3.2 p. 15) is a communication primitive according to which two messages ordered as per Lamport's "happened-before" relation (which in itself is often given the name *causality*) will be received in the same order by all the processes. This does not contradict the fact that this becomes a consistency criterion, i.e. a property of the objects implemented in the system: from the point of view of the program that uses the causally consistent shared object, the object is a part of the system.

Causal consistency is a way of weakening sequential consistency. The goal of sequential consistency is to provide a total order on all the operations of a concurrent history. This total order fixes a shared temporal reference, which is a reference for all the events of all processes. Causal consistency weakens sequential consistency by allowing time to be a relative concept for each process: the order which links all the events is only a partial order, called *causal order*. Like the total order of sequential consistency, causal order is not imposed by external conditions. All that is required is the *existence* of a causal order to explain concurrent histories. In particular, the causal order is not necessarily unique.

We will now formally define what a causal order is (definition 4.1). A causal order is a partial order that contains the sequential order $\mapsto$ of each process and, such that the causal future of an event is cofinite (its complement is finite). This last point means that no event can be indefinitely ignored by a process. This corresponds to the hypothesis that in the end, messages always arrive in the shared systems. There are three reasons why this is important in our model:

1) Without this restriction, the causal order could be the process order which is, in itself, a partial order. The criteria obtained will thus be much weaker as nothing will force the processes to interact. Similar criteria will thus be much weaker than local consistency and they could thus be implemented trivially, each process updating and reading its own local copy. However, such objects would be of no use in shared systems.

2) It is common to say that causal consistency is stronger than pipelined consistency. Based on the definition for pipelined consistency (p. 46), all the updates must appear in the linearization of all the processes. This property is the equivalent of the hypothesis on causal order.

3) We also look to define causal convergence, according to which convergence must be attained when all the processes have the same events in their causal past. In order for such a criterion to be stronger than eventual consistency, all the processes must end up having all the writes in their causal past.

The supplementary constraints verified by causal order will define the different consistency criteria.

DEFINITION 4.1.– Let $H = (\Sigma, E, \Lambda, \mapsto)$ be a concurrent history. A *causal order* is a partial order $\dashrightarrow$ on the events of E_H, which contains $\mapsto$ and such that, for any $e \in E_H$, $\{e' \in E_H : e \not\dashrightarrow e'\}$ is finite.

The set of causal orders for H is denoted by $co(H)$.

COMMENT 4.1.– We will next consider a history $H^{\dashrightarrow}$, which contains all the same events as H, but in which the process order has been replaced by the causal order $\dashrightarrow$. It is important to note that, for this, the causal order must be a well partial order (see p. 13). It is easy to show that any order relation that contains a well partial order is also a well partial order. As the process order is a well partial order, the causal order is also a well partial order.

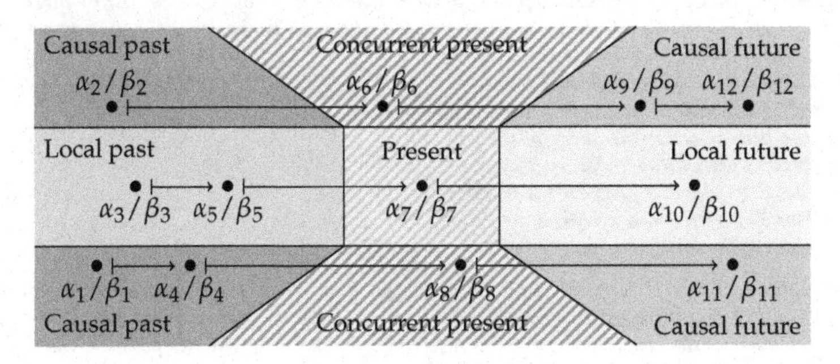

Figure 4.3. *Temporal zones defined by the causal order. For the color version of this figure, see www.iste.co.uk/perrin/distributed.zip*

A concurrent history that has a causal order is described by the two partial orders $\mapsto$ and $\dashrightarrow$. For each element e, each partial order for the set of events defines a temporal cone (similar to a light cone in special relativity) which may be separated into four zones: the past which contains the predecessors of e, the present, reduced to e, the future which contains the successors to e and the exterior of the cone which contains all the events incomparable to e. As $\mapsto \subset \dashrightarrow$, the cone local to the process is nested within that described by the causality. From the point of view of e, the concurrent history is thus partitioned into six zones represented in Figure 4.3:

– the *present* $\{e\}$;

– the *local past* $\lfloor e \rfloor_{\mapsto} \setminus \{e\}$, which corresponds to the past based on the strict process order;

– the *local future* $\lceil e \rceil_{\mapsto} \setminus \{e\}$, symmetrical to the local past in the future;

– the strict *causal past* $\lfloor e \rfloor_{\dashrightarrow} \setminus \lfloor e \rfloor_{\mapsto}$, that part of the causal past which is not already contained in the local past nor in the present;

– the strict *causal future* $\lceil e \rceil_{\dashrightarrow} \setminus \lceil e \rceil_{\mapsto}$, the equivalent of the strict causal past in the future;

– the *concurrent present* $E \setminus (\lfloor e \rfloor_{\dashrightarrow} \cup \lceil e \rceil_{\dashrightarrow})$, outside both cones.

The manner in which the present is affected by these different zones describes different consistency criteria.

WEAK CAUSAL CONSISTENCY

✗ Not composable p. 42
✓ Decomposable p. 42
✓ Weak p. 142

Weak causal consistency [PER 16b] ensures that each query returns a value in accordance with a complete knowledge of its causal past and only its causal past. Thus, in the history given in figure 4.2h (repeated here), the write $w(3)$ *of the third process is necessarily in the causal past of the second process when it reads* $(3, 1)$ *— from where else could we get the 3 that it returns? Additionally, as causal order must contain process order, the query of the second process precedes its update and, by transitivity, the write of 3 precedes the write of 2. This is why the two processes converge towards* $(3, 2)$ *and not towards* $(2, 3)$.

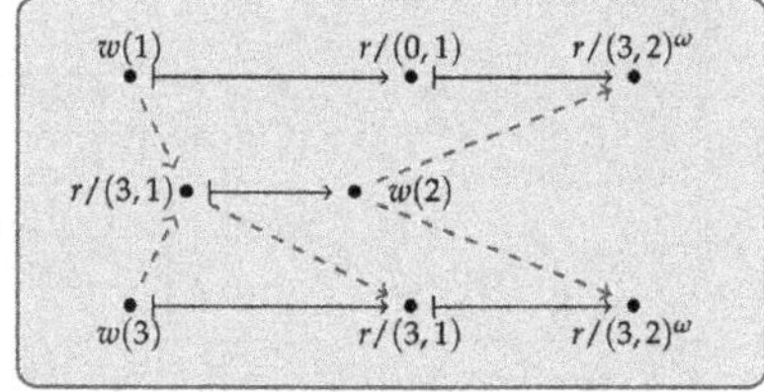

Formally, a history H *is weakly causally consistent with respect to an abstract data type* T *(definition 4.2) if there exists a causal order* $\dashrightarrow$ *such that for every event* e*, there exists a linearization of the causal past of* e *which leads to a state in which* e *is possible, that is a linearization of the history* $H^{\dashrightarrow}[\lfloor e \rfloor_{\dashrightarrow} / \{e\}]$*, which contains* e *and the queries in its causal past ordered according to the causal order. In the above example, causal order is represented by all the arrows (plain and dashed lines) and the following linearizations correspond to these criteria for the first queries of the three processes:*

$w(1) \cdot r/(0,1)$ $\qquad w(3) \cdot w(1) \cdot r/(3,1)$ $\qquad w(3) \cdot w(1) \cdot r \cdot r/(3,1)$

DEFINITION 4.2.– Weak causal consistency *is the consistency criterion*

$$WCC : \begin{cases} T \rightarrow & P(H) \\ T \mapsto & \left\{ H \in H : \begin{array}{l} \exists \dashrightarrow \in co(H), \\ \forall e \in E_H, \text{lin}\,(H^{\dashrightarrow}[\lfloor e \rfloor_{\dashrightarrow} / \{e\}]) \cap L(T) \neq \emptyset \end{array} \right\} \end{cases}$$

The history in figure 4.2d (to the right) is not weakly causally consistent. In fact, the read $(0, 1)$ *of the second process has the write of 1 in its causal past and the write of 2 in its causal future. We thus have* $w(1) \dashrightarrow w(2)$. *However, the write of 2 must precede that of 1 in the linearization of the read* $(2, 1)$*, which contradicts the causal order.*

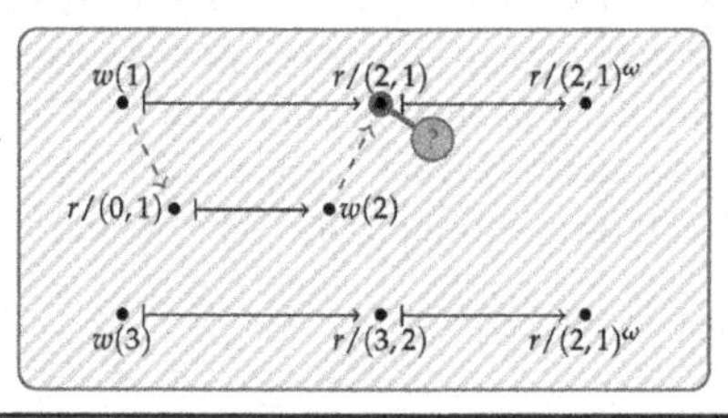

Figure 4.4. *Weak causal consistency*

4.3.2. *Weak causal consistency*

The existence of a sole causal order on a concurrent history H that verifies the definition 4.1 brings absolutely no information to this history, for two reasons:

1) Let us take the analogy of sequential consistency: it is not enough to totally order the events in a concurrent history to make them sequentially consistent; the linearization corresponding to the total order proposed must also satisfy the sequential specification of the object. This is the same for causal order: just as all order relations may be extended to total order, a causal order may be found for any concurrent history (any linear extension, for example, would qualify). The first question that comes up, then, is how causal order is related to the values that the operations may return, to give meaning to this causal order;

2) The dependency relationship between the reads and writes is a central idea in the original definition of causal memory. The second question is the manner in which the causal order is forced to contain these dependencies.

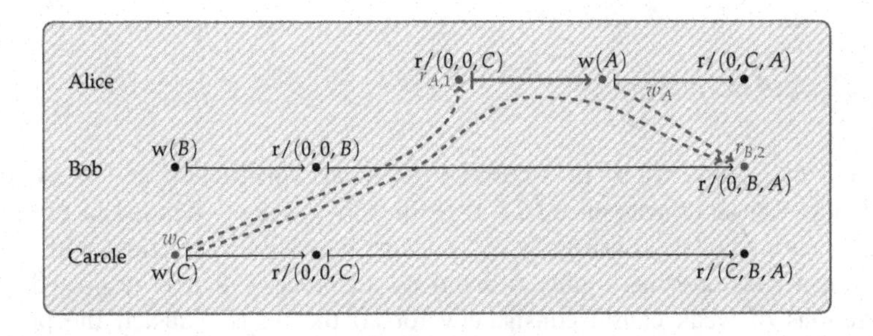

Figure 4.5. *Weak causal consistency applied to the history given in the preamble*

Weak causal consistency (WCC) [PER 16b], formally defined in Figure 4.4, responds to these two problems in the following manner: any read must be compatible with a state that results from the execution of all the writes in its causal past, and only these, in an order that respects the causal order.

This definition forces the causal order to contain the dependency relationships between the reads and writes in the case of a weakly casually consistent memory: if a process reads a value v from a register x, the sequential ordering of the writes in its causal past must lead to a state in which the value of x is v, but this is not possible unless the causal past contains a write for v in x. We will study the case of causal memory in more detail in section 4.4.2.

Let us now look at how weak causal consistency can be applied to the example given at the beginning of this chapter, that of Alice, Bob and Carole. Figure 4.5 models the history using a sliding window register, of size three. The sending of a message is modeled by the writing of a value A, B or C, and the reading of the message is a sliding window register. Let us try to construct a causal order $\dashrightarrow$ that verifies the required property. During the first read (event $r_{A,1}$), Alice sees Carole's message. Any linearization conforming to the sequential specification of the message queue that ends in this reading must necessarily contain the sending of the message by Carole (event w_C). As the writes for this linearization are exactly those in the causal past of the read, we have $w_C \dashrightarrow r_{A,1}$. Using the same reasoning for the message sent by Alice (event w_A) and the first read where Bob sees Alice's message (event $r_{B.2}$), we can deduce that $w_A \dashrightarrow r_{B,2}$. Additionally, we know, based on the definition 4.1 that the causal order contains the process order and, therefore, that Alice's first reading causally precedes the sending of her message. But then, $w_C \dashrightarrow r_{A,1} \dashrightarrow w_A \dashrightarrow r_{B,2}$; thus, the linearization required for the reading $r_{B,2}$ by Bob must necessarily contain the event w_C, which does not correspond to the observed reading. Weak causal consistency therefore allows for the specification of the desired behavior.

4.3.3. *Causal convergence*

There are many ways to reinforce eventual consistency by weak causal consistency. Causal convergence (CCv) [PER 16b], formally defined in Figure 4.6, reinforces strong update consistency by replacing the visibility relations with a causal order. The writes are thus totally ordered by an order $\leq$ that is respected by the linearizations of weak causal consistency for all the reads. Thus, if the processes create a finite number of writes, the number of reads that do not have all the writes in their cause past will also be finite. All other reads will have all the writes in their causal past, ordered in the same order. They will thus be executed in the same state.

The algorithm given in Figure 4.14 functions in the same way as the algorithm in Figure 3.7 as concerns strong update consistency. Process p_i controls a Lamport clock, $vtime_i$, which allows it to timestamp all operations so as to order them totally, and a variable, $history_i$, which contains all the history. In order to query the current state, p_i executes all the history beginning with the initial state. To execute an update operation α, it causally broadcasts α and the stamp that makes it possible to order it. Upon receiving such a message, p_i simply inserts the events in its variable $history_i$. This only differs from the algorithm in Figure 3.7 in the use of the causal broadcast, which makes it possible to ensure the transitivity of the visibility relation.

PROPOSITION 4.1.– *Any concurrent history allowed by the algorithm in Figure 4.7 is causally convergent.*

PROOF.– Let $T \in \mathcal{T}$ and H be a history admitted by the algorithm in Figure 4.7. Let $e, e' \in E_H$ be two events invoked by the processes p_i and $p_{i'}$ in the states $(vtime_i, history_i)$ and $(vtime_{i'}, history_{i'})$, respectively. We posit:

– $e \dashrightarrow e'$ if $e = e'$ or if the process executing e' has received the message sent during the execution of e' before executing e. As the reception is causal, $\dashrightarrow$ is indeed a causal order. As any message is received instantaneously by its emitter, $\mapsto\ \subset\ \dashrightarrow$. As any message is eventually delivered, $\dashrightarrow$ is a causal order;

– $e \leq e'$ if $vtime_i < vtime_{i'}$ or $vtime_i = vtime_{i'}$ and $i < i'$. This lexicographical order is total, as two operations from the same process have different clocks and the identifier for each process is unique. Additionally, if $e \dashrightarrow e'$, according to lines 9 and 13, $vtime_{i'} \leq vtime_i + 1$, thus: $e \leq e'$.

CAUSAL CONVERGENCE

✗ Not composable p. 42
✓ Decomposable p. 42
✓ Weak p. 142

Causal convergence [PER 16b] reinforces both strong update consistency as well as weak causal consistency. To do this, it reinforces strong update consistency by replacing the visibility relation by a causal order (that is: by imposing transitivity on the visibility relation). The history in figure 4.2g (given alongside the text) is causally convergent: the causal order is indicated in dashed lines and the total order of strong update consistency is represented by blue circles that are filled when the updates becomes visible.

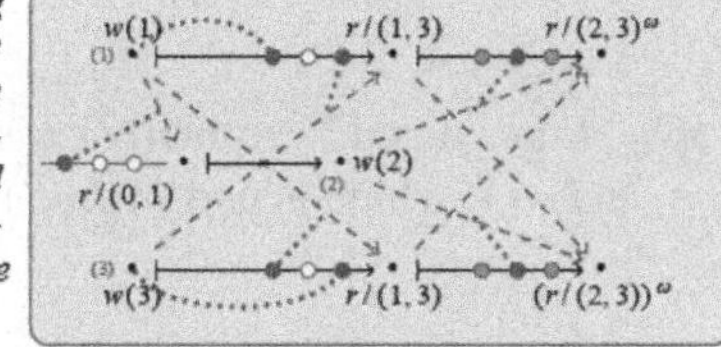

Formally, a history H is causally convergent with respect to an abstract data type T (definition 4.3) if there exists a causal order $\dashrightarrow$ and a total order $\leq$, which contains the causal order, such that each query results in the execution of the updates in its causal past, sorted according to the total order. In the below history, the updates are ordered $w(1) \leq w(2) \leq w(3)$ and one of the possible causal orders may be inferred based on the arrows formed by dotted lines. The queries correspond to the linearizations:

$w(1) \cdot r/(0,1)$ $\quad$ $w(1) \cdot w(3) \cdot r/(1,3)$ $\quad$ $w(1) \cdot w(2) \cdot w(3) \cdot r \cdot r/(2,3)$

DEFINITION 4.3.– Causal convergence *is the consistency criterion*

$$CCv : \begin{cases} \mathcal{T} \to \mathcal{P}(\mathcal{H}) \\ T \mapsto \left\{ H \in \mathcal{H} : \begin{array}{l} \exists \dashrightarrow \in co(H), \exists \leq \text{ total order on } E_H, \dashrightarrow \subset \leq \\ \wedge\ \forall e \in E_H, \text{lin}\left(H^{\leq}/\lfloor e \rfloor_{\dashrightarrow} / \{e\}\right) \cap L(T) \neq \emptyset \end{array} \right\} \end{cases}$$

The history given in figure 4.2h (opposite) is not causally convergent. In fact, in order to respect causal convergence, the update w(1) must causally precede the query r/(3,1) of the second process, which in itself must causally precede the update w(2). According to the total order, we must therefore have $w(1) \leq w(2)$. Furthermore, the reads of (1,3) impose that $w(3) \leq w(1)$. After convergence, the reads must return (2,3) and not (3,2).

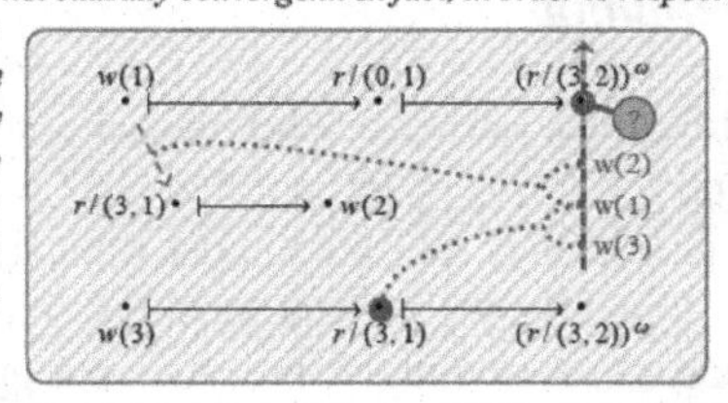

Figure 4.6. *Causal convergence*

```
1  algorithm CCv(A, B, Z, ζ0, τ, δ)
2      variable vtime_i ∈ ℕ ← 0;                                   // Linear clock
3      variable history_i ⊂ (ℕ × ℕ × A) ← ∅;                        // Known events
4      operation apply (α ∈ A) ∈ B
5          variable state ∈ Z ← ζ0;
6          for (t_j, j, α′) ∈ history_i sorted according to (t_j, j) do    // Read
7              state ← τ(state, α′);  // the history is played back
8          end
9          causal broadcast mUpdate (vtime_i + 1, i, α);            // Write
10         return δ(state, α);
11     end
12     on receive mUpdate (t_j ∈ ℕ, j ∈ ℕ, α ∈ A)
13         vtime_i ← max(vtime_i, t_j);
14         history_i ← history_i ∪ {(t_j, j, α)};
15     end
16 end
```

Figure 4.7. *Generic algorithm $CCv(T)$: code for p_i*

Let $e \in E_H$. Lines 6 and 7 explicitly construct a linearization of $H^{\leq}[\lfloor e \rfloor_{\dashrightarrow}/\{e\}]$ by defining $\leq$ and $\dashrightarrow$, which is part of $L(T)$ by defining T. □

4.4. Causal consistency

Causal memory [AHA 95] is stronger than PRAM [LIP 88]. In order for causal consistency to be an extension of causal memory for other abstract data types, it is essential that causal consistency be stronger than pipelined consistency. Weak causal consistency and pipelined consistency cannot be compared (Figures 4.2(d) and 4.2(g) give counterexamples in both directions). In this section, we present causal consistency [PER 16b], a new consistency criterion that is stronger than pipelined consistency as well as weak causal consistency.

4.4.1. *Definition*

The difference between weak causal consistency and pipelined consistency may be explained in terms of temporal zones, as shown in Figure 4.8. On the one hand, pipelined consistency seeks to impose a consistent view on each process during its life-span. Consequently, the present must take into account all its local past, both updates and queries. On the contrary, this does not guarantee the existence of a causal order, and therefore, the line between the update of the other processes that are considered and those that are ignored is blurred (Figure 4.8(a)). On the other hand, weak causal consistency seeks to force causal order and it thus requires that the present conforms to all the writes of the causal past (Figure 4.8(b)). Causal consistency (CC) reinforces weak causal consistency and pipelined consistency by considering the events of local past, as pipelined consistency does and the updates of the causal past (which are not part of the local past) like weak causal consistency does. To be more precise, for each event, there must be a linearization containing the writes of its causal past as well as the reads of its local past (Figure Figure 4.8(c)). The formal definition is given in Figure 4.9.

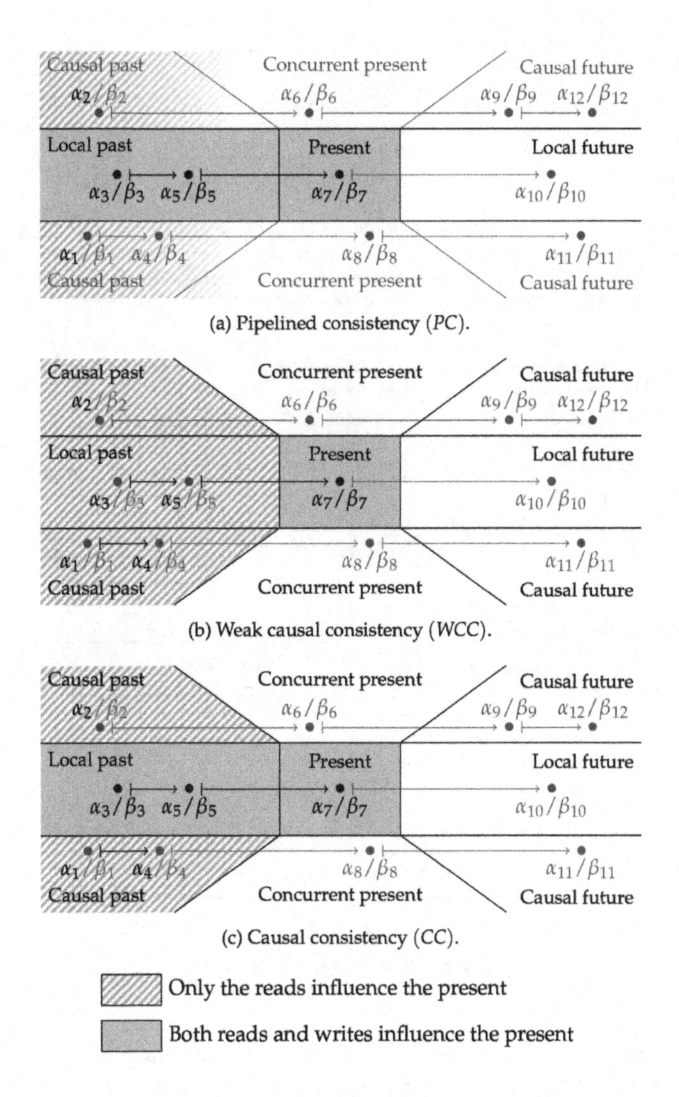

(a) Pipelined consistency (*PC*).

(b) Weak causal consistency (*WCC*).

(c) Causal consistency (*CC*).

Figure 4.8. *Consistency criteria using the six temporal zones. For the color version of this figure, see www.iste.co.uk/perrin/distributed.zip*

CAUSAL CONSISTENCY

✗ Not composable p. 42
✓ Decomposable p. 42
✓ Weak p. 142

Causal consistency [PER 16b] reinforces both weak causal consistency and pipelined consistency. Weak causal consistency imposes that each query returns a value with respect to its causal past and pipelined consistency prevents that the queries for the same process be independent of one another: in causal consistency, a query must, therefore, simultaneously take into account the updates in its causal past and the queries in its local past. For example, in the history in figure 4.2e (given alongside the text), the writes of 1 *and* 2 *are causally independent and, thus, it is equally likely that the last read of the first process returns* (2, 1) *as* (1, 2) *according to weak causal consistency. However, sequential specification forbids state* (2, 1) *from passing to state* (3, 1)*, from where we have the query* (1, 2) *in this history.*

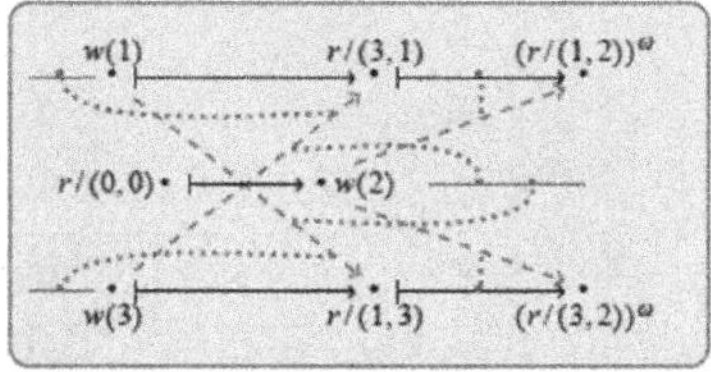

Formally, a history H is causally consistent with respect to an abstract data type T (definition 4.4) if there exists a causal order $\dashrightarrow$ such that for each event e, there exists a linearization of the updates in the causal past and the queries in the local past of e that leads to a state in which e is possible, that is, a linearization of the history $H^{\dashrightarrow}[\lfloor e\rfloor_{\dashrightarrow} / \lfloor e\rfloor_{\mapsto}]$, which contains the updates of the causal past $\lfloor e\rfloor_{\dashrightarrow}$ and the queries of the local past $\lfloor e\rfloor_{\mapsto}$ of e, ordered according to the causal order. In the above example, the following linearizations correspond to these criteria for the first two queries of the first process, where the silent values $\perp$ are left visible to differentiate the hidden updates from those that are not hidden:

$w(3)/\perp \cdot w(1) \cdot r/(3,1)$ $\qquad\qquad$ $w(3)/\perp \cdot w(1) \cdot r/(3,1) \cdot r \cdot w(2)/\perp \cdot r/(1,2)$

DEFINITION 4.4.– Causal consistency *is the consistency criterion*

$$CC: \begin{cases} T & \to \quad \mathcal{P}(H) \\ T & \mapsto \quad \left\{ H \in H : \begin{array}{l} \exists \dashrightarrow \in co(H), \\ \forall e \in E_H, \text{lin}\left(H^{\dashrightarrow}[\lfloor e\rfloor_{\dashrightarrow} / \lfloor e\rfloor_{\mapsto}]\right) \cap L(T) \neq \emptyset \end{array} \right\} \end{cases}$$

The history in figure 4.2i (to the right) is not causally consistent, even though it verifies both weak causal consistency and pipelined consistency. This is due to the fact that both writings of the value 3 *are interpreted differently by the two criteria. In pipelined consistency, linearization for the first process places the second write of* 3 *between the reads* (3, 2) *and* (2, 3)*, which creates a cycle in causal order between this write and the writing of* 1*. In weak causal consistency, the reads* (3, 2) *and* (2, 3) *have the same writes in their causal past, but the spontaneous change in state violates sequential specification. This shows that causal consistency is more than just the simple conjunction of weak causal consistency and pipelined consistency.*

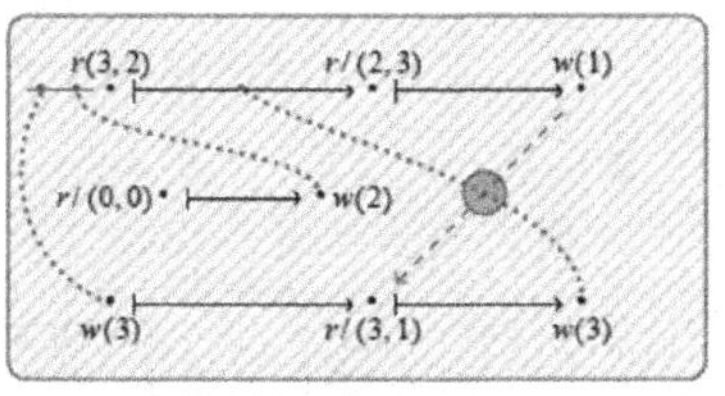

Figure 4.9. *Causal consistency*

Causal consistency is stronger than weak causal consistency because it only adds constraints. It is, on the other hand, not trivial that causal consistency is stronger than pipelined consistency (PC), given the definitions: the existence of a linearization for each event does not directly lead to one linearization for the whole history. We prove proposition 4.2, slightly stronger, which will be useful in section 4.4.2. The fact that $PC \leq CC$ is a direct corollary of this as $\mapsto \subset \dashrightarrow$..

WHAT IS A CAUSAL QUEUE?

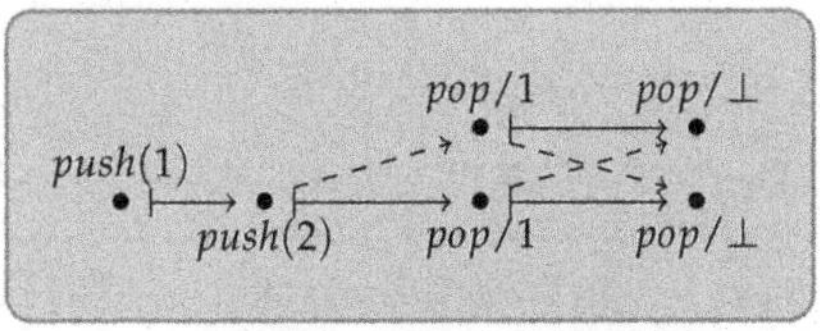

(a) The ADT queue $\mathcal{Q}_{\mathbb{N}}$

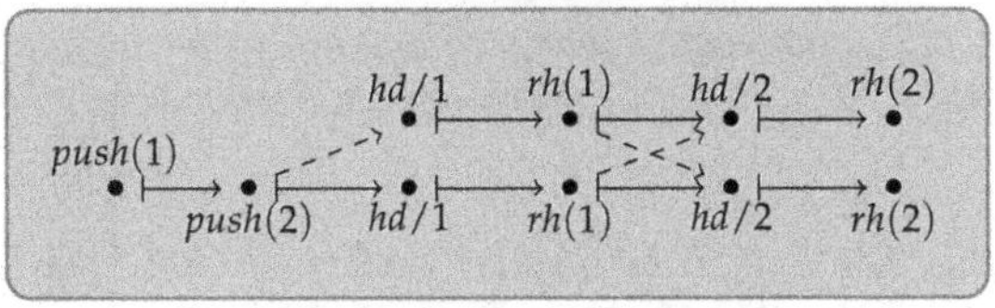

(b) The UQ-ADT queue $\mathcal{Q}'_{\mathbb{N}}$

The histories in figures 4.10a and 4.10b are both causally consistent, but not sequentially consistent. In the history given in figure 4.10a, when the two processes call the pop operation, they are in the same state $[1, 2]$ *and thus obtain* 1. *They then integrate the fact that the other process has deleted the head of the queue, which they believe is the remainder* 2. *During the next pop the queue is empty. No weak criterion can ensure that each element that is inserted will be taken once and only once if there are an infinite number of pop equations carried out. However, this example shows that causal consistency can guarantee neither the existence (2 is never read) nor unicity (1 is read twice) of a read for each element. This is because causal consistency – as well as all weak criteria presented in this book – weakens sequential consistency by assigning a different role to queries and updates. This excludes the atomicity of operations which are both queries and updates, such as the pop operation. In figure 4.10b, the pop operation has been divided into an hd query operation, which reads the head without touching it, and an rh update operation which deletes the queue head (type* Q'_N *definition 1.7). This history is similar to the preceding case and the two processes read1 and then carry out* $rh(1)$*. By doing this, they do not delete the* 2 *at the head of the queue. This technique makes it possible to guarantee that all insertions correspond to at least one read.*

Figure 4.10. *What is a causal queue?*

Causal consistency is more than the exact addition of weak causal consistency and pipelined consistency: the history in Figure 4.2(i), given again in Figure 4.11, is a counterexample. This history verifies pipelined consistency: Figure 4.11(a) shows how to obtain the linearizations for these two processes. It is also weakly causally consistent: Figure 4.11(b) depicts a possible causal order. However, this is not causally consistent as the second read of the first process is problematic. The linearization for this read must contain at least those events labeled $\mathrm{w}(3)$, $\mathrm{w}(2)$, $\mathrm{r}/(3,2)$, $\mathrm{w}(3)$ and $\mathrm{r}/(2,3)$ in this order. Thus, the last write of the third process must necessarily be a part of the causal past of this read. This is, however, impossible as it would create a cycle within the causal order. The problem does not arise with weak causal consistency as this allows the same writes for 2 and for 3 to be viewed in a different order between the first and second read of the first process.

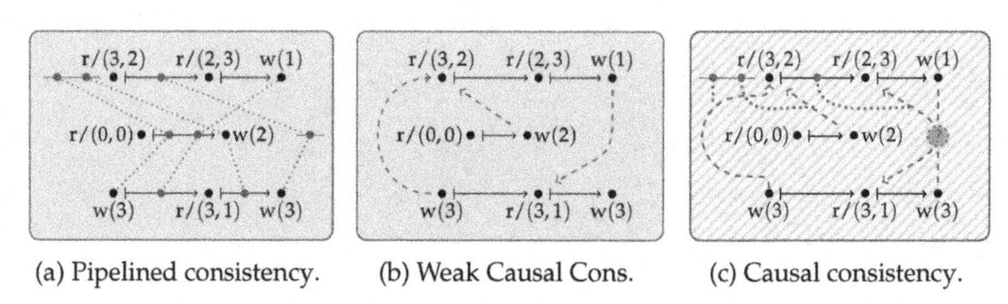

(a) Pipelined consistency. (b) Weak Causal Cons. (c) Causal consistency.

Figure 4.11. *Causal consistency is more than the combination of pipelined consistency and weak causal consistency*

PROPOSITION 4.2.– *Let $T \in \mathcal{T}$ be an ADT and $H \in CC(T)$ a causally coherent history. We have $\forall p \in \mathscr{P}_H, \text{lin}\,(H^{\dashrightarrow}[E_H/p]) \cap L(T) \neq \emptyset$.*

PROOF.– Let $T \in \mathcal{T}$, $H \in CC(T)$ and $p \in \mathscr{P}_H$.

If p is finite, it possesses a largest element e. As H is causally consistent, there exists a linearization $l_e \in \text{lin}(H^{\dashrightarrow}[\lfloor e \rfloor_{\dashrightarrow}/\lfloor e \rfloor_{\mapsto}]) \cap L(T)$. As e is maximal, $\lfloor e \rfloor_{\mapsto} = p$. As $\mapsto \subset \dashrightarrow$, there exists a linearization l for $H^{\dashrightarrow}[E_H/p]$ prefixed by l_e. As $l_e \in L(T)$ and all the events which are in l and not in l_e are hidden, $l \in L(T)$.

If not, p is infinite and does not possess a largest element. Instead, we construct an increasing set of linearizations $(l_n)_{n\in\mathbb{N}}$, which converges towards a sequential history that has the required properties. The linearizations given by causal consistency for successive events are not necessarily prefixed by one another (see Figure 4.14). The linearizations contain, therefore, in addition to the past of these events, a part of their concurrent present. The events of p are numbered in the order $e_1 \mapsto e_2 \mapsto \ldots$ and we define, for all $n \geq 1$, the set L_n such that $l.e_n \in L_n$ if and only if there exists an l', where $l.e_n.l' \in \text{lin}(H^{\dashrightarrow}[\{e_n\} \cup E_H \setminus \lceil e_n \rceil_{\dashrightarrow}/\lfloor e_n \rfloor_{\mapsto}]) \cap L(T)$. In other words, L_n contains the linearizations of the causal past and the concurrent present of e_n, truncated after e_n. L_n is not empty as it contains all the linearizations given by the causal consistency. It is also finite as $\dashrightarrow$ is a causal order, and thus, $E_H \setminus \lceil e \rceil_{\dashrightarrow}$ is finite. We also see that all the linearizations for L_{n+1} have a prefix within L_n because $e_n \dashrightarrow e_{n+1}$. $L(T)$ is therefore closed by taking the prefix.

We will now construct, by induction, a sequence $(l_n)_n$ of L_n words such that, for every n, l_n is a prefix of l_{n+1}. The empty chain $l_0 = \varepsilon$ is a prefix of a word in L_n for every n. Let us assume that for a given n, we have found a word $l_n \in L_n$, which is a prefix of a word in L_k word for every $k > n$. Let us show, by contradiction, that there exists a word in L_{n+1}, such that it has the prefix l_n. If this were not the case, as L_{n+1} is finite, there would be a k for which no word $l \in L_n$ would be a prefix of a word in L_k. As L_k is not empty, this then contradicts the fact that all words in L_k have a prefix in L_n. We have thus constructed a sequence $(l_n)_n$ of words from L_n such that, for every n, l_{n+1} has the prefix l_n. The series $(l_n)_n$ converges towards an infinite word l. As all the prefixes for l are within $L(T)$, $l \in L(T)$. Additionally, as $\dashrightarrow$ is a causal order, all the events are in the causal past of e_n for some n, and it follows that l contains all the events of $H[E_H/p]$. As the causal order is respected by all the prefixes for l, which contain two events, e and e', it is also respected within l. Finally, $l \in \text{lin}\,(H^{\dashrightarrow}[E_H/p]) \cap L(T)$. □

4.4.2. *Case study: causal memory*

It is important that the notion of causal consistency that we just defined be similar to that of causal memory, which we have already seen and which we will re-examine. The definition of causal memory is given on p. 50. In this section, we will compare causal consistency applied to memory and causal memory. We will first show that these two concepts are different by exhibiting a history accepted by causal memory but not causally consistent, in which the same value is written twice in the same register. We will then show that causal memory and causally consistent memory are identical when all the written values are different. Throughout this section, we will consider a countable set of register names, X, and a memory, $\mathcal{M}_X$, made up of these registers.

Causal memory explicitly defines the causal order by first considering a writes-into order, and then by using this causal order in the same manner as the process order in pipelined consistency. This approach reveals two limitations. First, there is no unicity in the writes-into order. Second, the construction of the linearizations for the processes is not highly restricted by the causal order. As a result of these two limitations, a read is not obliged to return a value written by its direct antecedent in the writes-into order. The history in Figure 4.12 illustrates this point. In this history, two processes write and read on a memory composed of four registers. The first process writes 1 then 2 in register a, 3 in b then reads 3 in d and 1 in c and finally again writes 1 in a. The second process behaves similarly by exchanging the roles of registers a and c and of registers b and d. This history verifies the property of causal memory: a possible writes-into order, drawn in straight lines ending in circles in Figure 4.12(a), links each read to the first write with the same value in the same register. The following linearizations, represented by the dotted lines, are correct:

$$w_a(1)/\bot.w_a(2)/\bot.w_b(3)/\bot.w_c(1).w_c(2).w_d(3).r_d/3.r_b.r_a.w_c(1).r_c/1.w_a(1)/\bot$$

$$w_a(1).w_a(2).w_b(3).w_c(1)/\bot.w_c(2)/\bot.w_d(3)/\bot.r_b/3.r_d.r_c.w_a(1).r_a/1.w_c(1)/\bot.$$

Therefore, the history verifies the property of causal memory. However, in these linearizations, the values read by the last two reads are not the result of their antecedents in the writes-into order. If we change this relation to re-establish the real data dependencies, we end up with a cycle in the causal order. The history is also not causally consistent, as seen in Figure 4.12(b). The write $w_b(3)$ necessarily precedes the read $r_b/3$ in the causal order and, therefore, the causal past of the read $r_a/1$ contains the writes $w_a(1)$ and $w_a(2)$ in that order. With respect to the sequential specification, the second write $w_a(1)$ must be placed between the write $w_a(2)$ and the read $r_a/1$ in the linearization of the latter and, thus, in the causal order. Symmetrically, the last write in the register c must causally precede the read $w_c(1)$, and this creates a cycle in the causal order.

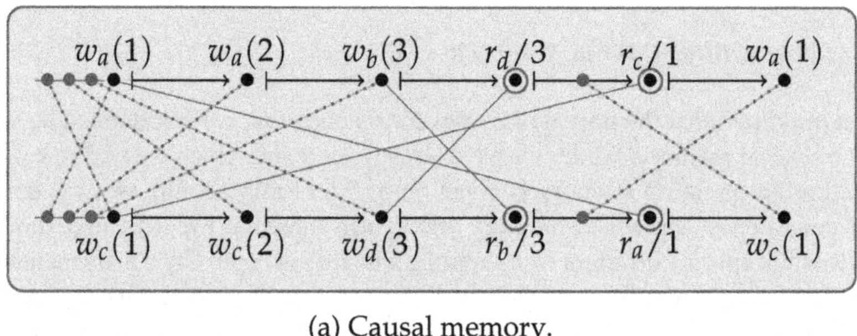

(a) Causal memory.

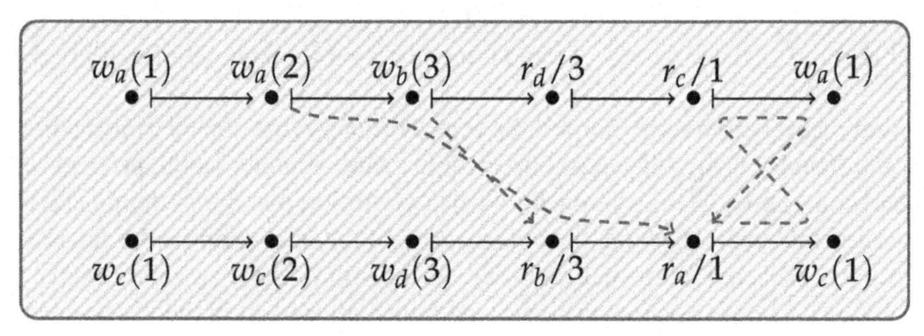

(b) Causally consistent memory.

Figure 4.12. *Causal consistency is different from causal memory*

This problem is a well-known limitation in causal memory [MIS 86], arising from the fact that the causal memory approach, which is based on operation semantics, is not well adapted for the definition of consistency criteria. This problem is usually resolved through the hypothesis that all the written values are distinct. This does not restrict the use of the memory as it is always possible to add information (e.g. a stamp similar to those used for strong update consistency) to the written value to make it unique. However, this solution is not ideal for specification as it artificially changes the operations given by the object and, thus, the sequential specification. We now show that, under the hypothesis of the uniqueness of written values, causally consistent memory and causal memory are identical. This means that our definition of causal consistency solves the above problem, by being equal to causal memory in all non-problematic cases.

PROPOSITION 4.3.– *Let H be a concurrent history. If H is causally consistent with respect to the memory on X ($H \in CC(\mathcal{M}_X)$), then H is $\mathcal{M}_X$-causal.*

PROOF.– Let us assume that H is causally consistent. For any event e, there exists a linearization $l_e \in \text{lin}(H^{\dashrightarrow}[\lfloor e \rfloor_{\dashrightarrow} / \lfloor e \rfloor_{\mapsto}]) \cap L(\mathcal{M}_X)$. We must note that this linearization is not necessarily unique, but we can fix it for each event going forward. We define the writes-into order $\multimap$ by $e \multimap e'$ if $\Lambda(e') = \text{r}_x/n$ and e is the event corresponding to the last write on x in l_e. As $l_e \in L(\mathcal{M}_X)$, $\Lambda(e) = \text{w}_x(n)$. e' has at most one antecedent by $\multimap$, and if it does not have any, $n = 0$.

The transitive closure $\xrightarrow{CM}$ of $\multimap \cup \mapsto$ is a partial order contained in $\dashrightarrow$. As per proposition 4.2, for any $p \in \mathscr{P}_H$, $\text{lin}\,(H^{\dashrightarrow}[E_H/p]) \cap L(\mathcal{M}_X) \neq \emptyset$, thus H is $\mathcal{M}_X$-causal. □

PROPOSITION 4.4.– *Let H be a concurrent history such that, for all $e, e' \in E_H$ where $\Lambda(e) = w_x(n)/\bot$ and $\Lambda(e') = w_y(p)/\bot$, $e \neq e' \Rightarrow (x, n) \neq (y, p)$. If H is $\mathcal{M}_X$-causal, then $H \in CC(\mathcal{M}_X)$.*

PROOF.– Let us assume that H is $\mathcal{M}_X$-causal. $\xrightarrow{CM}$ is a causal order. Let $p \in \mathscr{P}_H$ and $e \in p$. There exists a linearization $l_p \in \text{lin}(H^{\multimap CM}[E_H/p]) \cap L(\mathcal{M}_X)$, associated with a total order on the events $\leq_p$. Let l_e be the unique linearization of $\text{lin}(H^{\leq_p}[\lfloor e \rfloor_{\dashrightarrow} / \lfloor e \rfloor_{\mapsto}]$. Let $e' \in \lfloor e \rfloor_{\dashrightarrow}$ labeled r_x/n. If e' has no antecedents in the writes-into order, $n = 0$. If it exists, this antecedent e'' is the last write on x before e' in l_p as it is the unique event with the label $w_x(n)$ in H. As $e'' \xrightarrow{CM} e'$, it is also the last write on x before e' in l_e. Finally, $l_e \in L(\mathcal{M}_X)$, therefore H is causally consistent. □

4.4.3. *Implementation*

The problem of implementing causal memory has been widely studied in the literature [BAL 06, BAL 04, HÉL 06, MIL 06]. The generic algorithm given in Figure 4.14 extrapolates this work to all abstract data types. Each process p_i governs a local variable, $state_i$, which reflects the current state of the object. In order to execute an operation α, it carries out a query in this state and broadcasts the input symbol for the operation α using the causal broadcast primitive. When the process p_j receives the message broadcast by p_i, it locally applies the operation to its present state.

When it comes to memory, it is well known that this strategy implements a little more than causal memory [GAM 00]. Causally consistent situations which cannot be produced by the algorithms in Figure 4.13 are called *false causality situations*. Figure 4.14 explores this new concept by proposing a new consistency criterion, *strong causal consistency*, which accepts only the histories generated by the algorithm given in Figure 4.13. The history from Figure 4.2(e), detailed in Figure 4.14, is an example of false causality. The proposition 4.5 proves that strong causal consistency is the strongest consistency criterion implemented by the algorithm in Figure 4.13.

```
1  algorithm SCC(A, B, Z, ζ0, τ, δ)
2  |  variable state_i ∈ Z ← ζ0;                          // Local state
3  |  operation apply (α ∈ A) ∈ B
4  |  |  variable β ∈ B ← δ(state_i, α);                  // Read
5  |  |  causal broadcast message (α);                    // Write
6  |  |  return β;
7  |  end
8  |  on receive message (α ∈ A)
9  |  |  state_i ← τ(state'_i, α');
10 |  end
11 end
```

Figure 4.13. *Generic algorithm $SCC(T)$: code for p_i*

PROPOSITION 4.5.– *Consider an ADT $T \in \mathcal{T}$ and $H \in \mathcal{H}$, a concurrent history which represents the execution of a system formed of communicating processes (i.e. $\mathscr{P}_H$ forms a finite partition of E_H, each element $p \in \mathscr{P}_H$ being the set of events of one of the processes). H may be generated by the algorithm in Figure 4.13 if and only if $H \in SCC(T)$.*

PROOF.– Let $T \in \mathcal{T}$ and $H \in \mathcal{H}$, such that $\mathscr{P}_H$ forms a finite partition of E_H.

The principle underlying the algorithm in Figure 4.13 is to cause a local variable to locally evolve by applying to it all the writes when a process receives a message emitted during their call. The sequentiality of each process constructs a linearization that conforms to the linearization that is required for pipelined consistency corresponding to a total order$\dashrightarrow_p$. Additionally, the possible linearizations are restricted by the use of causal reception: if a process$p$ has already received an event e when it executes e' (according to $\dashrightarrow_p$), these two events are in the same order in the linearizations of all the processes. Thus, a concurrent history, H, is admitted by the algorithm in Figure 4.14, if and only if, for each process, there is a well partial order $\dashrightarrow_p$ of all the events of E_H, which contain the process order and such that:

– $\forall p \in \mathscr{P}_H, \mathrm{lin}\,(H^{\dashrightarrow_p}[E_H/p]) \cap L(T) \neq \emptyset$,

– $\forall p, p' \in \mathscr{P}_H, \forall e \in E_H, \forall e' \in p, (e \dashrightarrow_p e') \Rightarrow (e \dashrightarrow_{p'} e')$.

Let us assume that H is admitted by the algorithm in Figure 4.13. We posit that $\dashrightarrow\ = \left(\bigcap_{p \in \mathscr{P}_H} \dashrightarrow_p\right)$. As for all p $\dashrightarrow_p$ contains $\mapsto$, $\dashrightarrow$ also contains $\mapsto$. Let

$e \in E_H$. As $\dashrightarrow_p$ is a well partial order, $\lfloor e \rfloor_{\dashrightarrow_p}$ is finite, and thus, $\{e' \in E_H : e \not\dashrightarrow e'\} = \bigcup_{p \in \mathscr{P}_H}$, as a finite union of finite sets, is finite and $\dashrightarrow$ is a causal order. For any p and any $e \in p$, we define l_e as the prefix until e of the unique linearization for $\mathrm{lin}(H^{\dashrightarrow_p}[E_H/p]) \cap L(T)$. As $\dashrightarrow \subset \dashrightarrow_p$, $l_e \in \mathrm{lin}(H^{\dashrightarrow}[\lfloor e \rfloor_{\dashrightarrow}/\lfloor e \rfloor_{\mapsto}]) \cap L(T)$. Let $e, e' \in E_H$ such that $e \mapsto e'$. There exists a p such that $\{e, e'\} \subset p$; therefore, l_e and $l_{e'}$ are two prefixes of l_i, where l_e is shorter than $l_{e'}$, and therefore l_e is a suffix to $l_{e'}$. Finally, $H \in SCC(T)$.

STRONG CAUSAL CONSISTENCY

✗ Not composable p. 42
✓ Decomposable p. 42
✓ Weak p. 142

Strong causal consistency is the criterion that we obtain when all processes execute the updates of other processes in an order that is compatible with causal reception. The difference between strong causal consistency and causal consistency is that the successive linearization of events which take place in the same process are prefixed by one another. The history given in figure 4.2f (given alongside the text) is strongly causally consistent: for example, the first process writes 1 and passes to the state (0, 1), then executes the write of 3 and passes to the state (1, 3), which it reads, it then executes the write of 2 and finishes in the state (3, 2). The other processes behave in a similar manner.

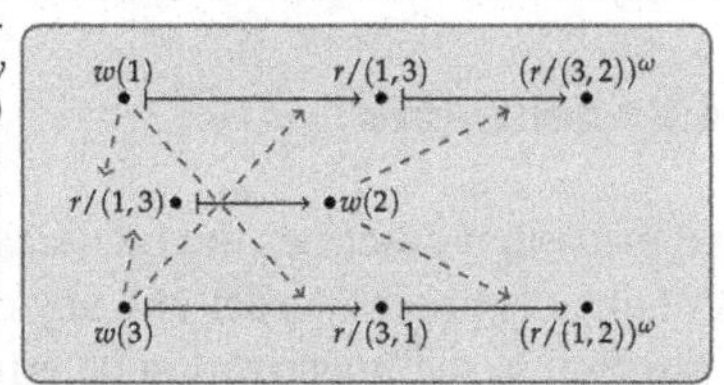

Formally, a history H is strongly causally consistent with respect to an abstract data type, T, (definition 4.5) if there exists a causal order $\dashrightarrow$ *and a sequence* l_e *for each event e in H such that :*

– *for any e,* l_e *is a linearization of the history* $H^{\dashrightarrow}[\lfloor e \rfloor_{\dashrightarrow}/\lfloor e \rfloor_{\mapsto}]$ *made up of queries of the local past and updates of the causal past of e, sorted according to the causal order (this is the property required by causal consistency) ;*

– *the linearizations of the events in a process are prefixed by one another: for any pair of events ordered according to the process order* $e \mapsto e'$, l_e *is a prefix of* $l_{e'}$.

In the example, the linearizations of the first three operations of the first process are:

$w(1)/\bot \qquad w(1)/\bot \cdot w(3) \cdot r \cdot r/(1,3) \qquad w(1)/\bot \cdot w(3) \cdot r \cdot r/(1,3) \cdot w(2) \cdot r/(3,2)$

DEFINITION 4.5.– Strong causal consistency *is the consistency criterion*

$$SCC : \begin{cases} \mathcal{T} & \rightarrow \quad \mathcal{P}(\mathcal{H}) \\ T & \mapsto \quad \left\{ H \in \mathcal{H} : \begin{array}{l} \exists \dashrightarrow \in co(H), \exists (l_e)_{e \in E_H}, \\ \forall e \in E_H, l_e \in \mathrm{lin}(H^{\dashrightarrow}[\lfloor e \rfloor_{\dashrightarrow}/\lfloor e \rfloor_{\mapsto}]) \cap L(T) \\ \wedge \ \forall e, e' \in E_H, e \mapsto e' \Rightarrow \exists l' \in \Sigma^\star, l_{e'} = l_e \cdot l' \end{array} \right\} \end{cases}$$

The history given in figure 4.2e (to the right) is causally consistent but not strongly causally consistent. In fact, by ignoring the hidden queries, the linearization for the first read of the first process can only be w(3) · w(1)/ ⊥ · r/(3, 1). This results in the fact that the linearization of the first update of the first process is w(3)·w(1)/⊥, and thus the write of 3 precedes that of 1 in the causal order. The situation is symmetrical with the third process, which results in a cycle in the causal order.

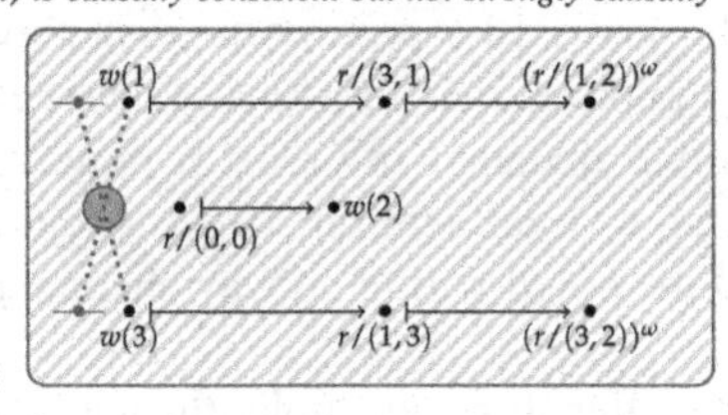

Figure 4.14. *Strong causal consistency*

Reciprocally, let us assume that $H \in SCC(T)$. We must note l_e, the linearization of each event, e, defined by strong causal consistency. Let us re-examine the construction of proposition 4.2. Let $p \in \mathscr{P}_H$. We begin by defining $\dashrightarrow'_p$ as: for all $e, e' \in E_H$, $e \dashrightarrow'_p e'$, if $e \dashrightarrow e'$ or if there exists $e'' \in p$ such that, in the linearization $l_{e''}$, e is placed before e' or e is present but not e'. The relation $e \dashrightarrow'_p e'$ is indeed an order as the linearizations of the events of p are related by a prefixation relationship (the events are placed in the same order) and they respect the causal order. We extrapolate $\dashrightarrow'_p$ to a total order $\dashrightarrow_p$. This contains $\mapsto$ and, by construction, the single linearization l_p of $\text{lin}(H^{\dashrightarrow_p}[E_H/p])$ contains l_e for any $e \in p$, and thus, $l_p \in L(T)$. Let $p, p' \in \mathscr{P}_H$, $e \in E_H$ and $e' \in p$ such that $e \dashrightarrow_p e'$. As e' appears in its own linearization, this means that e is a part of the linearization of e', and thus, $e \dashrightarrow e'$. This results in $e \dashrightarrow'_{p'} e'$ and, thus, $e \dashrightarrow_{p'} e'$. Finally, H is accepted by the algorithm in Figure 4.13. □

4.5. Specific behaviors

The total order that describes the linearization of sequential consistency is a causal order: it contains the process order and any event has a cofinite causal future because its (finite) position in the linearization determines its causal past. Additionally, as the order is total, the concurrent present of all the events is empty. The opposite is also true: a weakly causally consistent history in which each event has an empty concurrent present is sequentially consistent.

This means that it is not necessary to manage synchronizations in the same program several times over. More precisely, the histories generated by a distributed application using a shared object in its implementation are restricted both by the specification of the shared object (its sequential specification and its consistency criterion) and by the application itself (for instance, if an operation of the object is never called in the implementation of the application, this operation will not be present in the generated histories). If a program mechanism guarantees that two different operations are always causally dependent, a weakly causally consistent object behaves like its sequentially consistent counterpart. Let us imagine, for example, a game where each player plays by turn, such as chess. According to the rules of the game (the program), each player (a process) must wait for the other player to complete their turn (causal dependency between the displacement of two pieces, or two updates) in order to move their piece across the board (the shared object). Using weak causal consistency rather than sequential consistency to implement the board will in no way change the program behavior.

In fact, in the above example, the causal order is not total unless the chess clock rings between each turn (external synchronization mechanism). If not, the player who is not currently moving a piece (update) must look at the board (query) to know when they can play. The original article on causal memory [AHA 95] justifies the interest in this by the fact that any history in which two writes are always causally ordered is sequentially consistent. Proposition 4.6 proves that this is, in fact, a property of weak causal consistency: any weakly causally consistent history in which all the writes are ordered two by two is sequentially consistent (SC).

PROPOSITION 4.6.– *Let an ADT T and a concurrent history $H \in WCC(T)$. If, for any pair of updates $(u, u') \in U^2_{T,H}$, $u \dashrightarrow u'$ or $u' \dashrightarrow u$, then $H \in SC(T)$.*

PROOF.– Let $T \in \mathcal{T}$ and $H \in WCC(T)$ such that for any pair of updates $(u, u') \in U^2_{T,H}$, $u \dashrightarrow u'$ or $u' \dashrightarrow u$.

Let $\leq$ be a total order on E_H, which extends $\dashrightarrow$, and let l be the unique linearization of $\text{lin}(H^{\leq})$. As $\mapsto \subset \dashrightarrow \subset \leq$, $l \in \text{lin}(H)$. Let us assume that $l \notin L(T)$. As the transition system for T is deterministic, there exists a non-empty, finite prefix for l which does not belong to $L(T)$. Let $l' \in \Sigma^*$ and $e \in E_H$ such that $l' \cdot \Lambda(e)$ is the shortest such prefix. As $H \in WCC(T)$, there exists a linearization in $\text{lin}(H^{\dashrightarrow}[\lfloor e \rfloor_{\dashrightarrow}/\{e\}]) \cap L(T)$ of the form $l'' \cdot \Lambda(e)$ because e is the maximum for $\lfloor e \rfloor_{\dashrightarrow}$ according to $\dashrightarrow$. However, as l' and l'' have the same updates ordered in the same order, because $\dashrightarrow$ is total for the updates, l' and l'' lead to the same state. In these conditions, it is absurd to have $l'' \cdot \Lambda(e) \in L(T)$, $l' \in L(T)$ and $l' \cdot \Lambda(e) \notin L(T)$. Thus, $l \in L(T)$ and $H \in SC(T)$. □

A simple way to use this property (though one that has low tolerance for crashes) is to use a synchronization chip. A program implemented according to this strategy uses, in addition to a shared object representing the data, a chip whose value continuously indicates the identifier for a process. Only the process indicated by the chip is allowed to update the first object. When a process finishes its updates, it indicates another process by writing its identifier onto the chip. Such a strategy guarantees that two updates are always causally ordered. Proposition 4.6 thus allows for an affirmation that in this situation weak causal consistency, applied in the composition of the first object and the chip, offers exactly the same quality of service as sequential consistency, but at a much lower cost.

The proposition can also be applied to the other causal criteria discussed in this chapter, which are stronger than weak causal consistency. Additionally, proposition 4.7 proves that causally convergent histories, in which no update is incomparable with a query according to the causal order, are also sequentially consistent. Causal convergence reinforces update consistency,in which some queries may seem incorrect. This proposition affirms that the only queries that may appear incorrect in causal convergence are those which are incomparable with at least one update based on causal order.

PROPOSITION 4.7.– *Let there be an ADT T and a concurrent history $H \in CCv(T)$ such that for any $u \in U_{T,H}$ and $q \in Q_{T,H}$, $u \dashrightarrow q$ or $q \dashrightarrow u$. Thus, $H \in SC(T)$.*

PROOF.– Let $T \in \mathcal{T}$ and $H \in CCv(T)$ such that for any $u \in U_{T,H}$ and $q \in Q_{T,H}$, $u \dashrightarrow q$ or $q \dashrightarrow u$. As $H \in CCv(T)$, there exists a total order, $\leq$ that contains $\dashrightarrow$ and, for any $e \in E_H$, a linearization $l_e \cdot \Lambda(e) \in \text{lin}(H^{\leq}[\lfloor e \rfloor_{\dashrightarrow}/\{e\}] \cap L(T)$.

Let l be the unique linearization for $\text{lin}(H^{\leq})$. As $\mapsto \subset \leq$, $l \in \text{lin}(H)$. Let us assume that $l \notin L(T)$. As in the proposition 4.6, l has a prefix, $l' \cdot \Lambda(e) \notin L(T)$ where $l' \in L(T)$. The event e cannot be labeled a pure update as, when the transition system for T is complete, any pure update may take place in any state, and therefore, particularly in the state obtained after the execution of l'. The event e cannot, therefore, be a query. Thus, by the hypothesis, it is causally ordered with all the updates of that history. As the total order, $\leq$, respects the causal order, $\dashrightarrow$, the updates in the past of e, according to causal order and according to the total order, are exactly the same: $\lfloor e \rfloor_{\dashrightarrow} \cap U_{T,H} = \lfloor e \rfloor_{\leq} \cap U_{T,H}$. We can deduce from this that l_e and l' have the same updates in the same order and that it is thus impossible that $l_e \cdot \Lambda(e) \in L(T)$, $l' \in L(T)$ and $l' \cdot \Lambda(e) \notin L(T)$. Finally, $l \in L(T)$ and $H \in SC(T)$. □

4.6. Conclusion

In this chapter, we studied causality through consistency criteria. We extrapolated the concept of causal memory to all abstract data types by defining causal consistency as a consistency criterion. We also explored the variants of causal consistency around the four consistency criteria.

These four criteria are relevant. *Weak causal consistency* may be seen as the causal denominator of the other three causal criteria. *Causal convergence* is the result of the conjunction of weak causal consistency and update consistency. *Causal consistency*, which reinforces weak causal consistency and pipelined consistency, is a generalization of causal memory to all abstract data types. Finally, all abstract data types have an implementation that respects each of these criteria in wait-free systems. *Strong causal consistency* exactly models the standard implementation of causal consistency through message-passing systems.

Finally, this chapter allows for a deeper understanding of what causality brings to consistency criteria. The two criteria to keep in mind are causal consistency and causal convergence, which reinforce pipelined consistency and update consistency, respectively.

One question that comes up is that of effective implementation of causal convergence. Causal convergence corresponds to strong update consistency in which the visibility relation is transitive (hence, the similarity in the proofs for the algorithms in Figures 4.14 and 3.7). On the other hand, strong update consistency is more expensive to implement than update consistency. In particular, only the first of the three algorithms discussed in Chapter 3 implements strong update consistency and can, therefore, be adapted to implement causal convergence. However, the properties brought in by strong update consistency which are absent in the case of update consistency are not always necessary with weak causal consistency, which also specifies the queries carried out before convergence. Whenever this is possible, it can be cheaper and just as relevant to use a combination of weak causal consistency and update consistency, rather than causal convergence. It would be interesting to study higher-performing algorithms to implement the conjunction of weak causal consistency and update consistency.

5

Weak Consistency Space

5.1. Introduction

In this chapter, we will specifically study shared objects that can be implemented in wait-free message-passing asynchronous distributed systems (or, more simply, "wait-free systems", $AS_n[\emptyset]$, see section 1.3.2). According to the CAP theorem [BRE 00], there is no algorithm that can simultaneously guarantee all three of the following properties:

Strong consistency. In their proof for the CAP theorem [GIL 02], Gilbert and Lynch defined strong consistency using an atomic register (i.e. a linearizable one). The result may be extrapolated to sequentially consistent memory, on the condition that at least two registers are considered: in [LIP 88], Lipton and Sandberg prove that, in order to implement sequentially consistent memory in an asynchronous system, even without taking into consideration any crashes, the sum of the execution time for a write-operation and the execution time for a read-operation must be at least equal to the time taken to deliver a message. The result holds true for a wait-free system, which assumes fewer hypotheses.

Availability. Availability requires that the calls to an object method always eventually return a response. This corresponds to the progress condition we imposed to any algorithm in section 1.4.1

Partition tolerance. A *partitionable* system is subject to temporary or permanent *partitioning* during which two processes in different partitions cannot communicate with each other. In the proof for the CAP theorem [GIL 02], the partitions are modeled by loss of messages between processes in different partitions. If there is only a single process in one partition, it is impossible for this process to receive messages from any other processes. Wait-free systems are, therefore, good models for partitionable systems.

With respect to this discussion, we will denote the CAP theorem by theorem 5.1: it is impossible to implement a sequentially consistent memory that is made up of at least two registers in an asynchronous distributed system, where we do not suppose the hypothesis of a majority of correct processes. The result holds true for wait-free systems. The proof for this theorem is not given here as it is very similar to the end of the proof for theorem 5.7.

THEOREM 5.1.– *Let x and y be two register names, $n > 1$ processes and $f \geq \frac{n}{2}$. There is no algorithm that implements $SC\left(\mathcal{M}_{\{x,y\}}\right)$ in $AS_n[t \leq f]$.*

Problem. *Which shared objects may be implemented in wait-free systems?*

The first question that we ask ourselves is about the consistency criteria for which it is possible to implement *all* abstract data types (ADT) in a wait-free system. Such a consistency criterion is called a *weak criterion*, as opposed to *strong criteria* such as sequential consistency and linearizability that are subject to the CAP theorem. In the preceding chapters, we have identified update consistency and the variations around causal consistency as weak criteria. The objective of this chapter is to study the overall space of weak criteria. To do this, we follow the long quest for the strongest weak criteria.

Our second question concerns those objects that may be implemented in wait-free systems, verifying a given strong criterion. This will allow us to better understand the properties of abstract data types which make it possible to implement the strongest criteria.

Approach. *We are looking for pairs of weak criteria whose conjunctions results in a strong criterion. To prove that a consistency criterion is strong, we adapt the tools developed for distributed algorithms to wait-free systems.*

This approach allows us to demonstrate the absence of any strongest weak consistency criterion. To be more precise, we identify three weak criteria, said to be *primary* which may be conjugated in pairs to form *secondary* weak criteria, but whose conjunctions results in a strong criterion. The three secondary criteria obtained are pipelined consistency, update consistency and serializability. This makes it possible to explain the structure that can be seen in Figure 5.1(a).

For this, we prove that the Consensus number for a sequentially consistent sliding window register of size k, and with the conjunction of the three weak criteria, is equal to k. This makes it possible to elegantly fill Herlihy's hierarchy by showing a shared object that is intelligible on every echelon in the hierarchy. Additionally, this shows the existence of shared objects that verify a consistency criterion that is strictly weaker than sequential consistency and that have a Consensus number that is not 1.

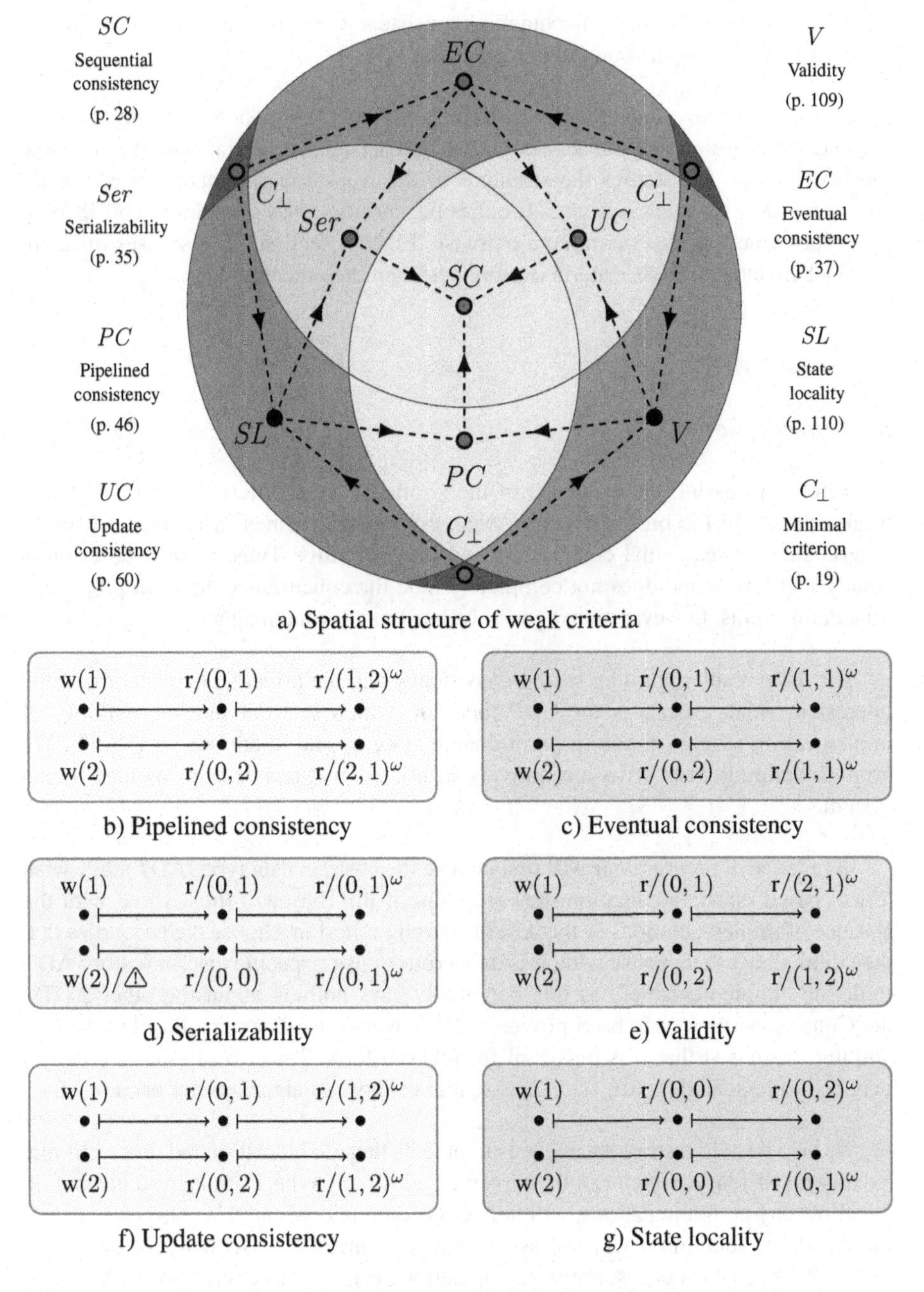

Figure 5.1. *The three paths of weak consistency applied to the object $\mathcal{W}_2$. For the color version of this figure, see www.iste.co.uk/perrin/distributed.zip*

Finally, we demonstrate the existence of a more precise hierarchy for abstract data types where the application of sequential consistency results in a Consensus number equal to 1 by varying the consistency criterion.

The rest of this chapter is organized as follows. Section 5.2 defines a weak consistency criterion and introduces two new consistency criteria: validity and state locality. Section 5.3 studies the structure of the weak criteria space by examining sliding window register. Section 5.4 studies the specific cases of memory and abstract data types where updates commute pairwise. Finally, section 5.5 discusses different cases where each of these criteria is more pertinent than the others.

5.2. Weak criteria

5.2.1. *Definitions*

Before discussing the existence of the strongest weak criterion, we must define weak criteria. In the broadest sense, weak criteria are defined in contrast to strong criteria such as sequential consistency and linearizability. Thus, a weak criterion is simply a criterion that does not completely hide the concurrence between processes. This definition is, however, too vague to be used in a formal manner.

The main result regarding strong consistency and the principal explanation of the interest in weak criteria is the CAP theorem, which sets out the impossibility of implementing strong criteria in partitionable systems and, therefore, in $AS_n[\emptyset]$. We propose defining weak criteria negatively, as criteria that *can* be implemented in this system.

In practice, a programmer will first choose the abstract data type (ADT) they wish to use based on the application they are using it in. They will then, faced with the absence of implementations or the cost of a strong criterion, choose the properties that they think they can dispense with. It is dangerous to use a specific discriminatory ADT to define "implementable", as this potentially says nothing about the other ADTs: the Consensus object has been proven to be universal for linearizability but there is nothing to suggest that it is universal for other criteria. Thus, to say that a criterion may be implemented in $AS_n[\emptyset]$, it is essential to show an algorithm for each ADT.

We also wish to restrict the analysis to objects that are indeed shared. For example, locality consistency, which can be implemented for any type, even in systems that do not allow any communications, will not be taken up in this study. We therefore define shared criteria (definition 5.1) as those which are comparable to sequential consistency (SC). The shared criteria space is partitioned between weak criteria, for which there is an algorithm that implements each ADT in wait-free systems (definition 5.2), and strong criteria (definition 5.3).

DEFINITION 5.1.– A criterion C is *shared* if $C \leq SC$ or $SC \leq C$.

DEFINITION 5.2.– A shared consistency criterion C is *weak* if, for any abstract data type T, there exists an algorithm A such that when it is executed in the system $AS_n[\emptyset]$,

– the algorithm A makes it possible to call all operations of T;

– all actions (operation calls and message reception) terminate;

– all the executions of A may be modeled by a C-consistent history for T.

The weak criteria space is designated by $\mathcal{C}_W$.

DEFINITION 5.3.– A shared consistency criterion is *strong* if it is not weak. The strong criteria space is designated by $\mathcal{C}_S$.

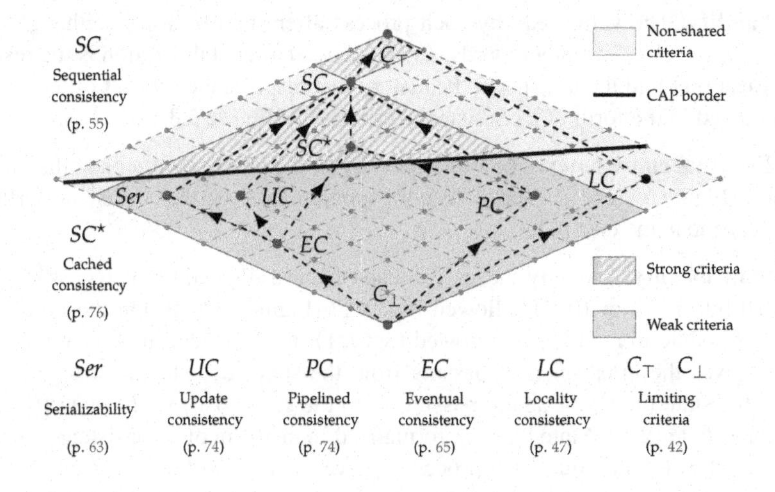

Figure 5.2. *Consistency criteria lattice. For the color version of this figure, see www.iste.co.uk/perrin/distributed.zip*

Based on these definitions, the existence of a strongest weak criterion seems quite unlikely. In fact, as represented in Figure 5.2, the set of shared criteria forms a cone with the apex SC. Moreover, the CAP theorem border, which separates strong criteria from weak criteria, cuts this cone below the apex. Therefore, $\mathcal{C}_W$ has the form of a truncated cone. In order for there to be a maximum in $\mathcal{C}_W$, it is essential that either the shared criteria cone be reduced to a line, that is, the shared criteria are totally ordered (at least locally, just below SC), or the CAP border be "parallel to one of the sides of the cone", that is, we should be able to identify one (or more) elementary properties for sequential consistency related to the CAP theorem. A shared criterion can be considered weak if, and only if, it does not guarantee this property.

5.2.2. *Validity and state locality*

In all the histories given in Figure 5.1, two processes share an instance of sliding window register of size 2. The first process calls the update $\mathrm{w}(1)$, then reads it an infinite number of times and the second process calls the update $\mathrm{w}(2)$, then reads it an infinite number of times. For most of the criteria studied in the preceding chapters, it was necessary that the sequence of values read by each process be consistent starting from a certain point, called the *stabilization* point. In these histories, the stabilization point is reached after the second read.

Let us first assume that the shared object is sequentially consistent. There exists $i \neq j$ such that $\{i, j\} = \{1, 2\}$ and the linearizations are part of the language $\mathrm{w}(i)/\bot \cdot \mathrm{r}/(0, i)^\star \cdot \mathrm{w}(j)/\bot \cdot \mathrm{r}/(i, j)^\omega$. The following three properties, in particular, are true:

Validity. The value read by each process after stabilization is either $(1, 2)$ or $(2, 1)$ (Figure 5.1(e)). In other words, any value read after stabilization is the result of a linearization of all the updates in the history that respects the process order. Validity is formalized in the form of a consistency criterion in Figure 5.3.

Eventual consistency. The values read by all the processes after stabilization are identical (Figure 5.1(c)). This property corresponds exactly to the consistency criterion with the same name (see p. 63).

State locality. The only successive reads that are allowed for a process are $(0, 1)$ followed by $(1, 2)$, or $(0, 2)$ followed by $(2, 1)$ (Figure 5.1(g)). On the other hand, it is not possible to read $(1, 2)$ followed by $(2, 1)$, for instance. In other words, each process gives the impression of passing from the state $(0, 0)$ to the state $(0, 1)$ and then to the state $(1, 2)$, or, again, passing from the state $(0, 0)$ to the state $(0, 2)$ and then to the state $(2, 1)$. State locality, formalized in the form of a consistency criterion in 5.4, models the fact that each process behaves as if it possesses a local state that only evolves as a result of the local execution of the updates present in the history.

We can now sort the consistency criteria defined in the preceding chapters depending on the properties that they verify.

Eventual consistency, validity or state locality only. Weak causal consistency is a stronger weak criterion than validity. Strong eventual consistency is a stronger weak criterion than eventual consistency.

Validity and state locality. Pipelined consistency (Figure 5.1(f)), causal consistency and strong causal consistency reinforce both validity and state locality. Furthermore, all three are weak criteria as they are weaker than strong causal consistency, which may be implemented as explained in chapter 4.

Eventual consistency and state locality. Serializability (Figure 5.1(d)) is stronger than both eventual consistency (in the version used in this book, where pure queries cannot abort) and state locality. It is also a weak criterion as it is possible to get all the updates to abort without even communicating.

Eventual consistency and validity. Update consistency (Figure 5.1(f)), strong update consistency and causal convergence reinforce both eventual consistency and validity. These are certainly weak criteria, based on the causal convergence algorithm presented in chapter 4.

Eventual consistency, validity and state locality. Of all the criteria discussed in the preceding chapters, only sequential consistency (and $C_{\top}$) simultaneously verifies all three properties. However, it is a strong criterion, based on the CAP theorem.

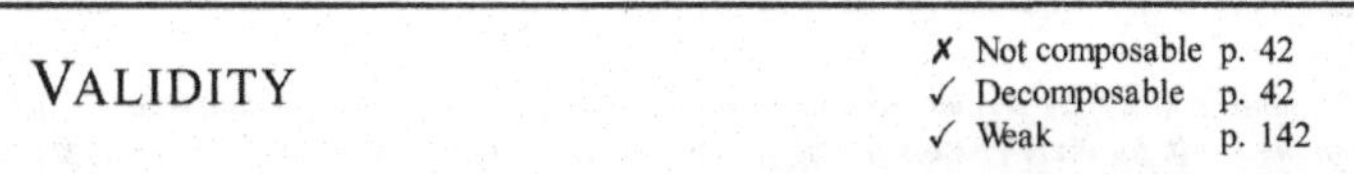

VALIDITY

✗ Not composable p. 42
✓ Decomposable p. 42
✓ Weak p. 142

Validity guarantees that if all the processes stop updating then, eventually, all the queried values will be compatible with a linearization of the updates in the history. For example, in the history in figure 5.1e (summarized alongside the text), the first process ends by reading the state (2, 1)*, which corresponds to the execution of* w(2) *followed by the execution of* w(1)*. The second process stabilizes itself in the state* (1, 2) *which corresponds to the execution of the two writes in the other direction.*

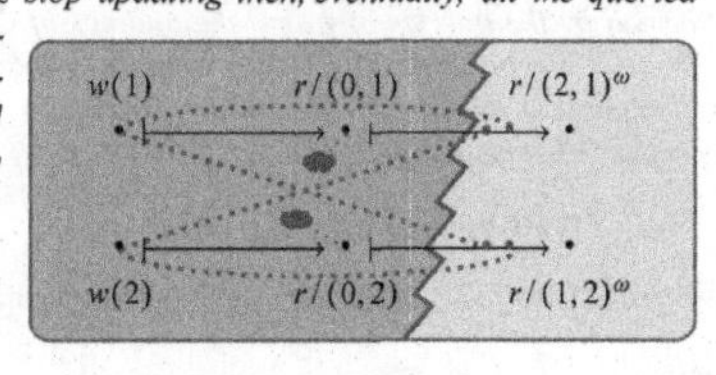

Formally, a history H is valid with respect to an abstract data type, T, (definition 5.4) if one of the two following properties is verified:

– *H contains an infinite number of updates* $(|U_{T,H}| = \infty)$

– *there exists a cofinite set of events, E′, such that for each of them, there exists a correct linearization with respect to the sequential specification of T of the history* $H[E_H/\{e\}]$*, which contains all the updates for H, as well as the event, e, being considered.*

In the below example, the set E′ contains the queries which return (2, 1) *or* (1, 2)*. The linearization required for the first query of each process is:*

$$w(2) \cdot r \cdot w(1) \cdot r/(2,1) \qquad w(1) \cdot r \cdot w(2) \cdot r/(1,2)$$

DEFINITION 5.4.– Validity *is the consistency criterion*

$$V : \left\{ \begin{array}{lll} T & \to & P(H) \\ T & \mapsto & \left\{ H \in \mathcal{H} : \begin{array}{l} |U_{T,H}| = \infty \\ \vee \quad \exists E' \subset E_H, (|E_H \setminus E'| < \infty \\ \quad \wedge \quad \forall e \in E', \mathrm{lin}\,(H[E_H/\{e\}]) \cap L(T) \neq \emptyset) \end{array} \right\} \end{array} \right.$$

The history in figure 5.1d (opposite) is not valid. In fact, the infinite number of queries of the second process are situated after the update w(2)*. Additionally, an infinite number of these updates must be present in E′. The only return values possible for these queries are* (1, 2) *and* (2, 1)*.*

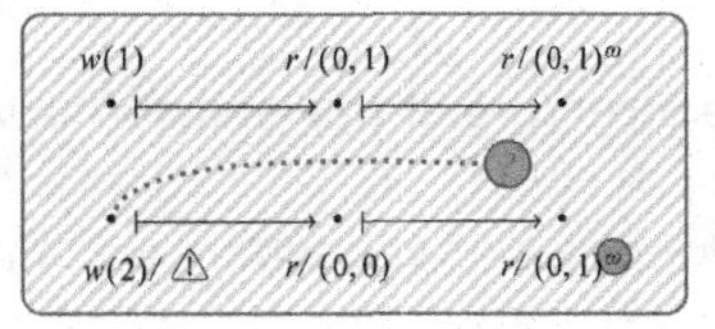

Figure 5.3. *Validity*

STATE LOCALITY

✗ Not composable p. 42
✓ Decomposable p. 42
✓ Weak p. 142

State locality models the fact that the state on a process cannot change spontaneously without executing an update of the history. This is translated by the fact that for every process, there exists a sequence of operations containing all the pure queries for the process and a part of the updates of the history, which is correct with respect to the sequential specification of the object: the state changes observed between two queries can, thus, only come from the execution of the updates situated between two queries in this linearization. In the history given in figure 5.1g (summarized alongside the text), the first process reads the value $(0,0)$ *then the value* $(0,2)$*. This can be interpreted as a passage from the local state* $(0,0)$ *to the local state* $(0,2)$ *during the execution of the update* $w(2)$ *carried out by the second process.*

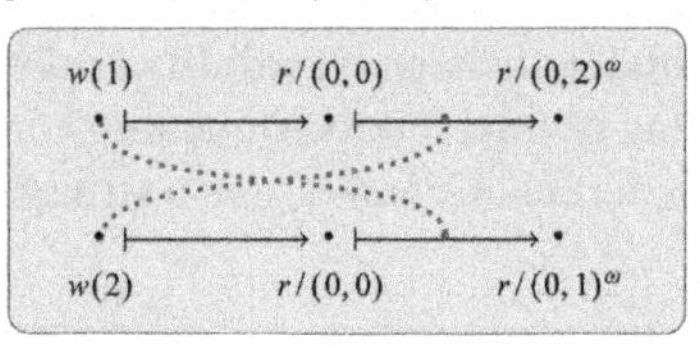

Formally, a history H verifies the state locality with respect to an abstract data type, T, (definition 5.5) if, for every process $p \in P_H$*, there exists a set of events,* C_p*, which contains all the pure queries (the projection only conserves those of p), such that the history* $H[p \cap C_p / C_p]$ *formed by the queries of p and the updates of* C_p *has a linearization compatible with the sequential specification for T. In the below example, the set* C_p *contains the pure queries and, for each process, the update of the other process. The two following linearizations correspond to the property for the two processes:*

$$r/(0,0) \cdot w(2) \cdot (r \cdot r/(0,2))^\omega \qquad\qquad r/(0,0) \cdot w(1) \cdot (r \cdot r/(0,1))^\omega$$

DEFINITION 5.5.– State locality *is the consistency criterion*

$$SL : \begin{cases} T & \rightarrow \quad P(H) \\ T & \mapsto \quad \left\{ H \in \mathcal{H} : \begin{array}{l} \forall p \in P_H, \exists C_p \subset E_H, \\ \hat{Q}_{T,H} \subset C_p \\ \wedge \ \ \mathrm{lin}(H[p \cap C_p / C_p]) \cap L(T) \neq \emptyset \end{array} \right\} \end{cases}$$

The history in figure 5.1f (to the right) does not verify state locality. In fact, in order for the second process to pass from the state $(0,2)$ *to* $(1,2)$*, it must execute an update* $w(1)$ *then an update* $w(2)$*. The history only contains one update* $w(2)$ *that the process has already executed to pass from the state* $(0,0)$ *to the state* $(0,2)$*. It is, thus, impossible to construct the wished-for linearization for the second process.*

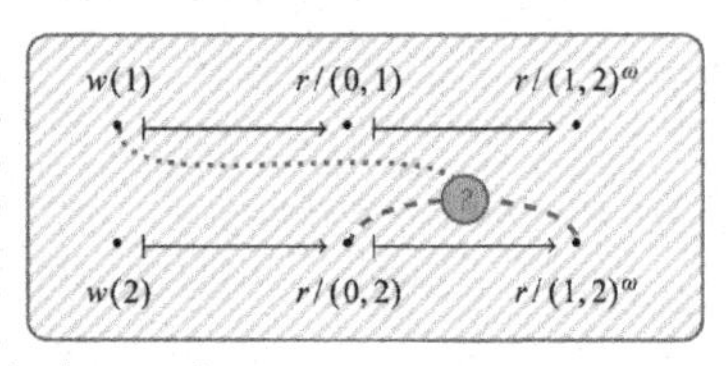

Figure 5.4. *State locality*

5.3. The structure of the weak criteria space

5.3.1. *Primary and secondary criteria*

Validity (V), eventual consistency (EC) and state locality (SL) are all weak criteria as they are all weaker than either causal convergence (validity and eventual consistency) or causal consistency (validity and state locality), for which we have given the algorithms in chapter 4. We will now prove that the conjunction of these three criteria $(V + EC + SL)$ is a strong criterion. For this, we will show that the sliding window register of size k $(V + EC + SL)$-consistent has a Consensus number equal to k, which means that it may be used to implement Consensus between k processes, but not more (see p. 58). As it is impossible to implement Consensus in wait-free systems [FIS 85], the sliding window register of size k is an example of the abstract data type for which $(V + EC + SL)$ cannot be implemented.

LEMMA 5.2.– *For any $k \in \mathbb{N}$, a sliding window register of size k $(V + EC + SL)$-consistent has a Consensus number greater than or equal to k.*

PROOF.– Let $k \in \mathbb{N}$. We will demonstrate that it is possible to implement Consensus in $AS_k[\emptyset]$ using a sliding window register of size k $(V + EC + SL)$-consistent.

We will consider the algorithm in Figure 5.5 to implement Consensus: to propose a value v, a process writes $v + 1$ in a sliding window register of size k $(V + EC + SL)$-consistent. Process p_i accesses this shared object through the local accessor $\mathbf{swr}_i$. Writing $v + 1$ rather than v makes it possible to ensure that all the written values are different from the default value, even if the proposed value is a null value. The process returns the first non-null value from its first read, from which it subtracts 1.

```
1 algorithm Consensus
2     variable swr_i : W_k;                    // swr_i verifies V+EC+SL
3     operation proposes (v ∈ N) ∈ N
4         swr_i.w(v+1);
5         variable tuple : N^k ← swr_i.r();
6         return tuple[min{0 ≤ j < k : tuple[j] ≠ 0}] − 1;
7     end
8 end
```

Figure 5.5. *Reduction of Consensus to a sliding window register $(V + EC + SL)$-consistent: code for p_i.*

We will now prove that this algorithm implements Consensus. For this, let us consider a history H produced by the execution of the algorithm. Let H' be a history obtained by adding an infinite number of readings following the writing and reading of all the correct processes during the execution of the algorithm. Each process p_i is in one of the three following situations:

– there is crash before the algorithm starts;

– it writes the value $v_i + 1$, then crashes;

– it writes the value $v_i + 1$, then reads the next values $\left(\zeta_j^i\right)_{j\in\mathbb{N}}$.

This history is represented in Figure 5.6. The information brought in by eventual consistency is in gray, that brought in by validity is in blue and that brought in by state locality is in red. If none of the processes are correct, the Consensus is trivially reached. We will, thus, suppose that at least one process is correct.

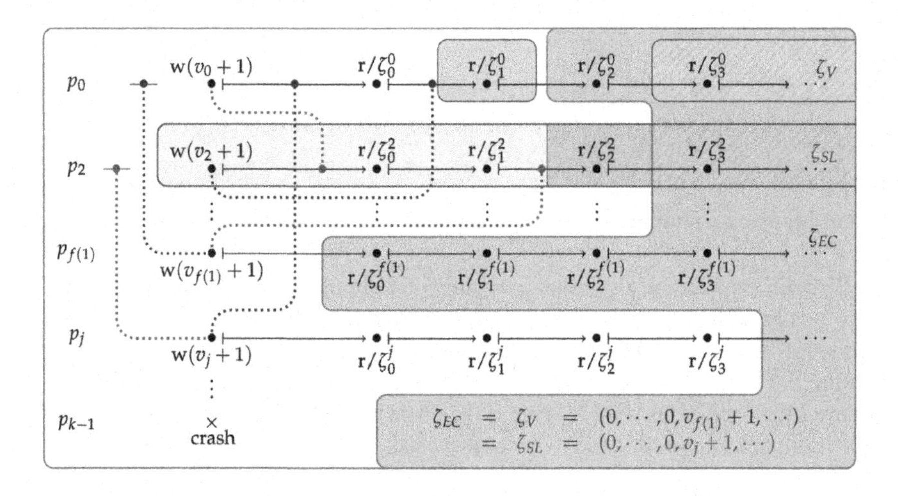

Figure 5.6. *Extension of the history produced by the algorithm in Figure 5.5. For the color version of this figure, see www.iste.co.uk/perrin/distributed.zip*

Eventual consistency. Because the shared object is eventually consistent, as the number of updates, $|U_{T,H}| \leq k$, is finite, there exists a state ζ_{EC} and a cofinite set of reads Q_{EC}, such that all reads of Q_{EC} return ζ_{EC}.

Validity. Let Q_V be the cofinite set of queries required for validity.

The sets Q_{EC} and Q_V are both cofinite with respect to E_H, and thus, there exists a query $e \in Q_{EC} \cap Q_V$, called by the process p_i. Based on validity, there exists a linearization $L_V \in \text{lin}(H[E_H/\{e\}]) \cap L(\mathcal{W}_k)$.

Let $f(j)$ denote the identifier for the process which has called the jth update of the linearization L_V. As L_V respects the sequential specification of the sliding window register, e returns the value $\zeta_V = (0, \cdots, 0, v_{f(1)} + 1, \cdots, v_{f(l)} + 1)$ formed by $k - |U_{T,H}|$ times the default value, 0 followed by $l \leq k$ update values. As $e \in Q_{EC}$, we have $\zeta_{EC} = \zeta_V$.

State locality. As the sliding window register verifies state locality, for each correct process p_i, there exists a set of events C^i_{SL} which contain all the pure queries in the history and a linearization L^i_{SL} which contains the queries for $C^i_{SL} \cap p_i$ and the updates for C^i_{SL}. Let Q^i_{SL} be the set of queries of p_i which appear after the last update in L^i_{SL}. Again, once $C^i_{SL} \cap p_i$ is cofinite, $(C^i_{SL} \cap p_i) \cap Q_{EC}$ is not empty. So there exists a query which returns $\zeta_{SL} = \zeta_{EC} = \zeta_V$. As ζ_V contains $|U_{T,H}|$ non-null values and ζ_{SL} contains $|C^i_{SL} \cap U_{T,H}|$ non-null values, we have $C^i_{SL} \cap U_{T,H} = U_{T,H}$: linearization L^i_{SL} contains all the updates in the history. We have now arrived at two deductions:

– As the update for p_i is part of C^i_{SL}, the first query for p_i does not return the initial state.

– As the size, k, of the sliding window register is greater than or equal to the number of updates, the first non-null value is the same in all the queries for p_i.

As the first non-null value for the queries for Q^i_{SL} is $v_{f(1)} + 1$, the first non-null value for the first query for p_i is also $v_{f(1)} + 1$, which is not dependent on i.

Let us go back to the algorithm. The history H, produced by the algorithm, is a prefix of H'. Any correct process returns $v_{f(1)}$ (termination), which is the common value returned by all the processes (agreement). Additionally, $v_{f(1)}$ is the value proposed by $p_{f(1)}$ (validity). Thus, the algorithm in Figure 5.5 does implement Consensus in $AS_k[\emptyset]$, and therefore, the sliding window register of size k $(V + EC + SL)$-consistent has a Consensus number greater than or equal to k. □

LEMMA 5.3.– *For any $k \in \mathbb{N}$, the sequentially consistent sliding window register of size k has a Consensus number less than or equal to k.*

PROOF.– This demonstration returns to the structure of proofs of the original article on the Consensus number [HER 91], which were themselves inspired by the demonstration of the impossibility of Consensus in message-passing asynchronous distributed systems where at least one outright crash may take place [FIS 85].

The central concept in these demonstrations is *valency*: a general state during the execution of an algorithm is *v-valent* if, from this global state, the algorithm can only return v, regardless of the scheduling of processes due to asynchronism. A global v-valent state for any value v is said to be *monovalent*. A global state that is not monovalent is said to be *plurivalent*. A process can only end in a monovalent global state. The idea, therefore, is to construct an infinite execution in which the global state remains plurivalent.

Let us assume the existence of an algorithm that resolves Consensus in an asynchronous system made up of $k + 1$ processes that communicate only through the intermediary of a sequentially consistent sliding window register of size k. Initially, all the correct processes call the **propose** method. We will assume that at least two different values are proposed.

Let p_i and p_j be two processes that propose different values, v_i and v_j. In an execution where all the processes except for p_i crash before they can call a sliding window register operation, p_i must return v_i to respect validity. Thus, v_i may be returned by the algorithm. Symmetrically, v_j may be returned by the algorithm. The global state is, thus, plurivalent.

Let us assume the existence of a global plurivalent state s, such that the call for the next sliding window register operation by any process results in a monovalent state. One of the two possible situations may follow depending on whether or not the next operation for a process is a query:

– Let us assume that there exists a process p_i whose next call is a query. The query leads to a global v_i-valent state, s'. As s is plurivalent, there exists a process p_j whose next call leads to a v_j-valent state, where $v_i \neq v_j$. Thus, there exists an execution from the state s, in which p_i does not carry out any call and p_j finishes by returning v_j. The global states s and s' differ only by the state locality for p_i, during the same execution from the state s', p_j also finishes by returning v_j. This contradicts the fact that s' is v_i-valent.

– In the other situation, the next call to all the processes is an update. As the global state is plurivalent, there exist two different values v_i and v_j and two processes p_i and p_j, such that the execution of p_i leads to a v_i-valent global state, s_i, and the execution of p_j leads to a v_j -valent global state, s_j. On the one hand, the successive call for update by p_j and then by the $k-1$ other processes leads to a state s'_j, which is v_j-valent because s_j is. There is, therefore, an execution from the state s'_j, in which p_i does not carry out any call and p_j finishes by returning v_j. On the other hand, the successive call by p_i, then by p_j, and then by the $k-1$ other processes for their update leads to the state s_i , which is v_i-valent because s_i is. The state of the sliding window register in s'_i and s'_j is the same because the last k updates are the same (that of p_j followed by those the $k-1$ other processes). As the global states s'_i and s'_j only differ by the state locality for p_i, during the same execution from the state s'_i, p_j also finishes by returning v_j. This contradicts the fact that s'_i is v_i-valent.

As such a state does not exist, it is possible to construct an infinite sequence of calls to sliding window register operations that all lead to plurivalent global states. This contradicts the fact that algorithms must terminate and, therefore, such an algorithm does not exist. It is impossible to implement Consensus for $k+1$ processes with a sequentially consistent sliding window register of size k, thus the Consensus number for the sequentially consistent sliding window register of size k is less than or equal to k. □

THEOREM 5.4.– *For any $k \in \mathbb{N}$, the shared objects $(V + EC + SL)(\mathcal{W}_k)$ and $SC(\mathcal{W}_k)$ have a Consensus number equal to k.*

PROOF.– Let $k \in \mathbb{N}$. We know that sequential consistency is stronger than the three primary criteria. It is, therefore, stronger than their conjunction. From this, we can deduce that the Consensus number for $(V + EC + SL)(\mathcal{W}_k)$ is less than or equal to that of $SC(\mathcal{W}_k)$. We also know that according to lemma 5.2, the Consensus number for $(V+EC+SL)(\mathcal{W}_k)$ is at least k and based on lemma 5.3, the Consensus number for $SC(\mathcal{W}_k)$ is, at most, k. Finally, the two shared objects have a Consensus number equal to k. □

COROLLARY 5.5.– *The conjunction of the three primary criteria* $(V+EC+SL)$ *is a strong criterion.*

PROOF.– Let $n > 1$ and $k \geq 2$. We know that, according to theorem 5.4, the Consensus number for the object $(V+EC+SL)(\mathcal{W}_k)$ is equal to k. From this, we can deduce that, based on[HER 91], there is no algorithm that implements $(V+EC+SL)(\mathcal{W}_k)$ in $AS_n[t \leq 1]$. Consequently, there is no such algorithm in $AS_n[\emptyset]$ either and thus, $(V+EC+SL)$ is a strong consistency criterion. □

Lemma 5.2 shows that the three consistency criteria studied are related to the three Consensus properties: validity makes it possible to affirm that the value returned by the Consensus algorithm will be one of the proposed values (Consensus *validity*), as it is obtained by linearization of the history updates; eventual consistency enables *agreement* as all the processes must result in the same value, and state locality ensures that the first non-null value in the queries will always be the same. This is a key point used to understand when an algorithm may stop and, thus, guarantee the *termination* of the Consensus. The corollary 5.5 makes it possible to divide the weak criteria space into six broad families contained between the minimal criterion $C_\perp$ and the family of strong criteria. On the one hand, eventual consistency, validity and state locality are three very weak properties, each of which corresponds to a Consensus property. These *primary criteria* may be conjugated two by two to form three new families of *secondary criteria*, each of which guarantees two of the three Consensus properties. Additionally, each family of secondary criteria has a family of complementary primary criteria, corresponding to the Consensus property that they do not guarantee. The conjunction of criteria from complementary families is a strong criterion. This division justifies the representation with primary and secondary colors, which is shown in Figure 5.1(a). We also note that two criteria from different secondary criteria families (respectively primary) are incomparable between themselves. If this were not the case, then, based on the lattice structure, their conjunction would be the strongest of the two, that is, a weak criterion (respectively primary). However, this is a strong criterion (respectively secondary).

Figure 5.7 represents a piece of the transition system for an abstract data type, T. The states for T are divided into three subsets: T_0, T_a and T_b, such that there is no transition between the states of T_a and the states of T_b. At a given moment when all the processes are in a state in T_0, two write operations a and b are carried out, such that executing a leads to a state in T_a and executing b leads to a state in T_b. Validity forbids the states from resting indefinitely in T_0, but it is impossible to guarantee, in wait-free systems, that the processes will all go into T_a or all into T_b. In the contrary case, they either remain separate and the history is not eventually consistent, or some must jump from one state to the other at the cost of state locality. All abstract data types that have this clause in their transition system are subject to this impossibility result. This is, in particular, the case for instant messaging services, which gives us the three strategies seen in the introduction: Hangouts remains in T_0 by sacrificing validity, WhatsApp gives up eventual consistency by presenting messages in the order in which they are received and Skype compromises on state locality by reordering messages. For sliding window registers, the size k indicates the minimal number of updates required to return to a common state from the states ζ_a and ζ_b. For example, for a sliding window register with size 2, we can arrive at the state $(3, 4)$ after two updates, beginning from the state $(0, 1)$ or the state $(0, 2)$. The history examined in lemma 5.2 contains at most k updates. The first are used to pass from T_0 to T_a and T_b and the ensuing states are not enough to return to a common state. We can note that as k increases, the transition system of the sliding window register of size k approaches that shown in Figure 5.7, which has an infinite Consensus number (if at least one query operation makes it possible to distinguish states from T_a and T_b), while the sliding window register has a Consensus number k.

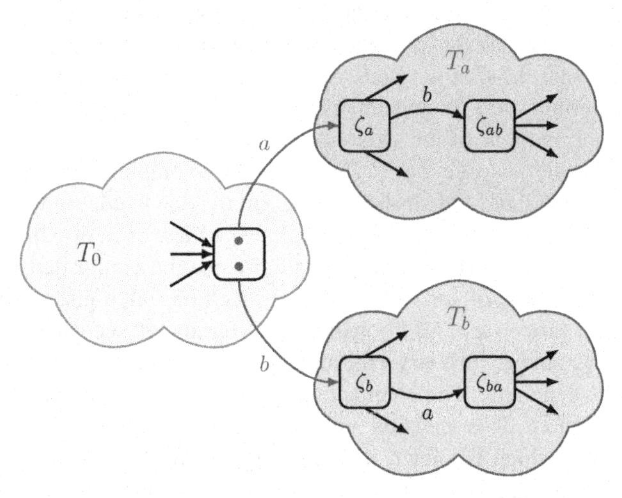

Figure 5.7. *Explanation of the structure of weak criteria space. For the color version of this figure, see www.iste.co.uk/perrin/distributed.zip*

5.3.2. *Plurality of decomposition*

A question that comes up naturally in the context of the decomposition of primary and secondary criteria is that of unicity. We will show that there is no unicity in this decomposition. For this, we will construct an infinite, uncountable family of weak consistency criteria, which, conjugated two by two, result in a strong criteria.

The aim of this study is to question the limitations of our model so as to respond to questions on the whole of weak criteria space. The question of countability is the first example of the questions that come up. The construction of theorem 5.6 will no longer work if we require the existence of a generic algorithm to implement all abstract data types. This is because such an algorithm will be obliged to calculate the binary representation of any real number, which is not generally possible. We can, however, show in the same way a *countable* infinity of weak criteria, which, conjugated two by two, results in a strong criterion. In addition, like a kaleidoscope, each of these criteria presents facets of the same consistency criteria (eventual consistency and pipelined consistency) seen from a different angle. Eventual consistency and pipelined consistency seem more "homogenous" and, thus, more "natural" than the criteria that we construct here. Therefore, the chief purpose of theorem 5.6 is to demonstrate that the introduction of such concepts is a prerequisite for any analytical approach of weak criteria.

However, such an approach is required in order to respond to questions such as those concerning the existence of maximal weak criteria, which may be asked in the following way: does the border separating the weak criteria space from strong criteria space contain weak criteria? One way of responding to these questions may be to propose a logic that makes it possible to express consistency criteria.

THEOREM 5.6.– *There is an uncountable infinity of weak consistency criteria which, when conjugated two by two, is a strong criterion.*

PROOF.– Let $n > 1$ processes. We will construct an uncountable infinity of consistency criteria $(C_x)_{x\in[0;2^{-n+1}[}$, such that there is at least one ADT which cannot be implemented in $AS_n[t < 1]$ for the conjunction of any two of them.

Let $x \in [0; 2^{-n+1}[$ be a real number. Its binary representation[1] is a sequence of digits $(x_k)_{k\geq 2}$, such that $x = \sum_{k=2}^{\infty} x_k 2^{-k}$. For any x, we associate the consistency criterion C_x, which behaves like pipelined coherence for a sliding window register whose size corresponds to 1 in the binary representation of x, and to eventual consistency for all other types:

$$C_x : \begin{cases} \mathcal{W}_k \mapsto PC(\mathcal{W}_k) \text{ if } x_k = 1 \\ T \mapsto EC(T) \text{ else} \end{cases}$$

The application $x \to C_x$ is injective as two different real numbers always have at least one different digit in their binary representation. In addition, $[0; 2^{-n+1}[$ is uncountable; therefore, $\{C_x : x \in [0; 2^{-n+1}[\}$ is uncountable too. For all x, C_x is indeed a weak criterion because for any T, $C_x(T) \in \{PC(T), EC(T)\}$. Thus, we

1 This representation is not unique (e.g. $0,001111... = 0,010000...$), but it is sufficient that it exists and we assign one to each x now.

now only have to show that the conjunction of C_x, two by two, is a strong criterion. Let x and y be two different real numbers from $]0; 2^{-n+1}[$. There exists $k \geq n$ such that $x_k \neq y_k$. Moreover, $(C_x + C_y)(\mathcal{K}_k) = (PC + EC)(\mathcal{K}_k)$, which does not have implementations in $AS_n[t < 1]$, based on theorem 5.4. From this, we deduce that $C_x + C_y$ is a strong criterion. □

5.4. Hierarchy of abstract data types

The preceding section proves that it is not possible to implement *all* ADTs with the conjunction of the three primary criteria. This does not mean that there is no implementation for *any* type. For example, we know that a unique, sequentially consistent register may be implemented in $AS_n[\emptyset]$, like a sliding window register of size 1, which behaves in the same way.

The study of consistency criteria gives us a new tool with which to study calculability in distributed systems. Thus, computability in distributed systems may be described as the study of a space with three "dimensions": systems, abstract data types and consistency criteria. The main results from this field, such as [ATT 95, HER 91] concern linearizability, and the variations in consistency criteria have never been studied with this objective, to the best of our knowledge. In this section, we will specifically study the strong consistency criteria for which certain abstract data types (memory and commutative abstract data types) may be implemented in wait-free systems. We show that this makes it possible to draw out a hierarchy between these abstract data types based on the strongest criterion that they can verify in a wait-free system.

5.4.1. *Memory*

It is actually possible to simultaneously implement eventual consistency, state locality and validity for a memory in wait-free systems: we can show that the algorithm in Figure 3.7 (p. 65) implements state locality and, therefore, the conjunction of the three weak criteria, as $SUC + SL > V + EC + SL$. This does not challenge the conclusion from the previous result and the division into families of primary and secondary criteria. Theorem 5.7 proves that, for memory, pipelined consistency (PC) and eventual consistency (EC) cannot be implemented together.

THEOREM 5.7.– *Let x, y and z be three register names, $n > 1$ processes and $f \geq \frac{n}{2}$. There is no algorithm that implements $(PC + EC)\left(\mathcal{M}_{\{x,y,z\}}\right)$ in $AS_n[t \leq f]$.*

PROOF.– Let x, y and z be three register names, $n > 1$ processes and $f \geq \frac{n}{2}$. Let us assume that there exists an algorithm, A, which implements $(PC + EC)\left(\mathcal{M}_{\{x,y,z\}}\right)$ in $AS_n[t \leq f]$. We are going to show (prove by contradiction) that there exists an execution that is allowed by A that does not respect the criterion $(PC + EC)$.

We can partition the set of processes into two subsets $\Pi_0 = \{p_0, ..., p_{f-1}\}$ and $\Pi_f = \{p_f, ..., p_{n-1}\}$, whose size is at least equal to $n - f$. We consider a program in which p_0 writes 1 in x then 1 in z, then reads, in turn x, y and z in an infinite loop; p_f writes 1 in y then 2 in z, then also reads the three variables indefinitely; the other processes do not read or write any variable. All the processes execute the algorithm A. This program generates the history given in Figure 5.8.

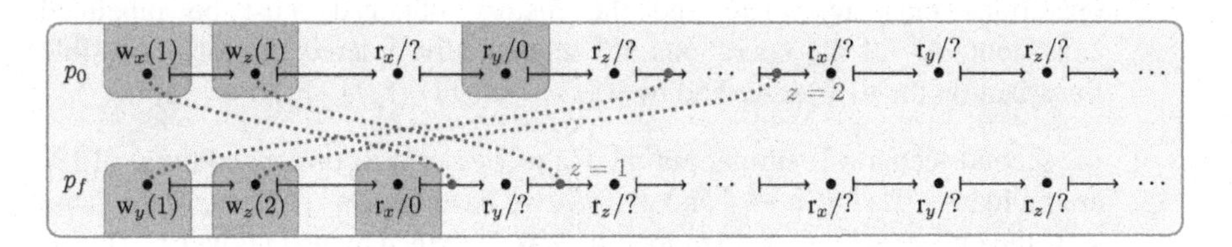

Figure 5.8. *Can we implement an eventually consistent memory and pipelined consistency?*

Is it possible that p_0 and p_f read $y = 0$ and $x = 0$, respectively, during their first read?

If this is possible, then, as A implements pipelined consistency, there must be a linearization of the history of p_0 containing all the updates and an infinite number of queries for each variable. The first query of y, which returns 0, must be placed before the update $w_y(1)$, which itself must be placed before the update $w_z(2)$, so as to respect the process order. The set of possible linearizations for p_0 is, therefore, the ω-regular language L_0 defined below. In the same way, the language for the possible linearizations of p_f is the language L_f:

$$
\begin{aligned}
L_0 &= w_x(1) \cdot w_z(1) \cdot (r_x/1 \cdot r_y/0 \cdot r_z/1)^+ \cdot w_y(1) \cdot (r_x/1 \cdot r_y/1 \cdot r_z/1)^\star \\
&\quad \cdot w_z(2) \cdot (r_x/1 \cdot r_y/1 \cdot r_z/2)^\omega \\
L_f &= w_y(1) \cdot w_z(2) \cdot (r_x/0 \cdot r_y/1 \cdot r_z/2)^+ \cdot w_x(1) \cdot (r_x/1 \cdot r_y/1 \cdot r_z/2)^\star \\
&\quad \cdot w_z(1) \cdot (r_x/1 \cdot r_y/1 \cdot r_z/1)^\omega
\end{aligned}
$$

The history must, thus, necessarily contain an infinite number of events labeled $r_z/2$ and an infinite number of events labeled $r_z/1$. Furthermore, no state in $\mathcal{M}_{\{x,y,z\}}$ accepts these two queries and the history contains only a finite number of updates. Eventual consistency cannot, therefore, be assured. This means that A must forbid p_0 from reading $r_y/0$ or forbid p_f from reading $r_x/0$.

However, this requirement is exactly what cannot be guaranteed, according to [ATT 94] and [GIL 02]. Let us consider the following three scenarios:

S_0 : in the first scenario, all the Π_0 processes crash from the beginning of the execution, before even sending a message, while those of Π_f are correct. Given that A is capable of tolerating up to f faults and $|\Pi_0| = f$, all the p_f operations must terminate and the history obtained must be pipelined consistent. As all the operations are sequentially ordered, the only possible linearization for p_f is described by $\mathrm{w}_x(1) \cdot \mathrm{w}_z(1) \cdot (\mathrm{r}_x/1 \cdot \mathrm{r}_y/0 \cdot \mathrm{r}_z/1)^\omega$.

S_f : the second scenario is similar but the Π_0 processes are correct and those of Π_f are faulty. As $|\Pi_f| = n - f$ and $f > \frac{n}{2}$, we have $|\Pi_f| \leq f$. Thus, in the same way, the only possible linearization for p_0 is described by $\mathrm{w}_y(1) \cdot \mathrm{w}_z(2) \cdot (\mathrm{r}_x/0 \cdot \mathrm{r}_y/1 \cdot \mathrm{r}_z/2)^\omega$.

$S_\emptyset$: in the last scenario, all the processes are correct, but the messages broadcast between the Π_0 and Π_f processes are different. From the point of view of p_0, during the first read of y, everything happens in exactly the same way as in S_f: there is no way of differentiating between a faulty process and a process where all the sent messages are very slow. Process p_0 must, therefore, behave as in S_f and return 0 during this read. Similarly, p_f, being incapable of differentiating this case from S_0, must return 0 during the first read of x.

Finally, for this program, algorithm A must both forbid the read $y = 0$ by p_0 or the read $x = 0$ by p_f to guarantee $PC + EC$, and authorize them to tolerate faults. As these two requirements are contradictory, A cannot exist. □

This result must make us wary of a common simplification of pipelined consistency, where this criterion, which respects process order, is a link between causal consistency (which respects causal order) and eventual consistency (which does not respect any order). On the one hand, validity and state locality are two important properties. On the other hand, it is wrong to think that pipelined consistency and causal consistency reinforce eventual consistency and any attempt to reinforce them by adding eventual consistency to their definition can only be made at the cost of losing one or the other primary criteria that make them up, or by exiting the wait-free model.

5.4.2. *Commutative data types*

Just as corollary 5.5 does not mean that no data type can respect all three primary criteria, theorem 5.7 does not say that pipelined consistency and eventual consistency can never be attained at the same time. In particular, in a case where all the operations commute, eventual consistency is very easy to obtain. This is the intuition that led to the development of CRDTs (commutative replicated data types). [SHA 11a]. As more complex objects whose operations do not commute (e.g. OR-Set) are also called CRDT [SHA 11b], we introduce the new name *CADT* (for "Commutative Abstract Data Types") to designate ADTs whose update operations are commutative. Among the CADTs we find, for example, the unbounded counter that is equipped with incrementation and decrementation operations or the set in which elements can only be inserted.

DEFINITION 5.6.– A *CADT* is an ADT $(\mathsf{A}, \mathsf{B}, \mathsf{Z}, \zeta_0, \tau, \delta)$, such that, for any pair of operations, $(\alpha, \alpha') \in \mathsf{A}^2$, and for any state, $\zeta \in \mathsf{Z}$:

$$\tau(\tau(\zeta, \alpha), \alpha') = \tau(\tau(\zeta, \alpha'), \alpha).$$

We surmise that the hypothesis about a majority of correct processes is necessary (conjecture 5.8) and sufficient (conjecture 5.9) to implement CADTs. This means that, from an implementation point of view, CADTs and memory require the same hypotheses. As a result, CADTs, like memory, have a Consensus number equal to 1.

CONJECTURE 5.8.– *Let $n > 1$ processes and $f \geq \frac{n}{2}$. There exists a commutative abstract data type T, such that there is no algorithm that implements $SC(T)$ in $AS_n[t \leq f]$.*

CONJECTURE 5.9.– *Let T be a commutative abstract data type. There exists an algorithm that implements $SC(T)$ in $AS_n[t < \frac{n}{2}]$.*

The weak consistency criteria theory highlights a difference in calculability between memory and CADTs: the conjunction of pipelined consistency and eventual consistency may be implemented for CADTs but not for memory.

In the transition system in Figure 5.7, a CADT will verify $\zeta_{ab} = \zeta_{ba}$. The execution of a and b must, therefore, always lead to this common state, whether the path to this state passes through ζ_a or ζ_b. Rather than proving that CADTs may be implemented in wait-free systems by showing an algorithm, theorem 5.10 proves that eventual consistency is freely obtained for any history that verifies pipelined consistency. This property explains the interest shown in CRDTs.

COMMENT 5.1.– The hypothesis that process order is a well order is important in the proof of theorem 5.10. Without this hypothesis, the existence of an infinite antichain of queries created in the initial state, differing from the convergence state, would be technically possible, which would prevent the guarantee of eventual consistency.

THEOREM 5.10.– *For any CADT T, $PC(T) \subset EC(T)$.*

PROOF.– Let T be a CADT and $H \in PC(T)$. If H has an infinite number of updates, then $H \in EC(T)$. Let us assume that H has a finite number of updates, as given in Figure 5.9.

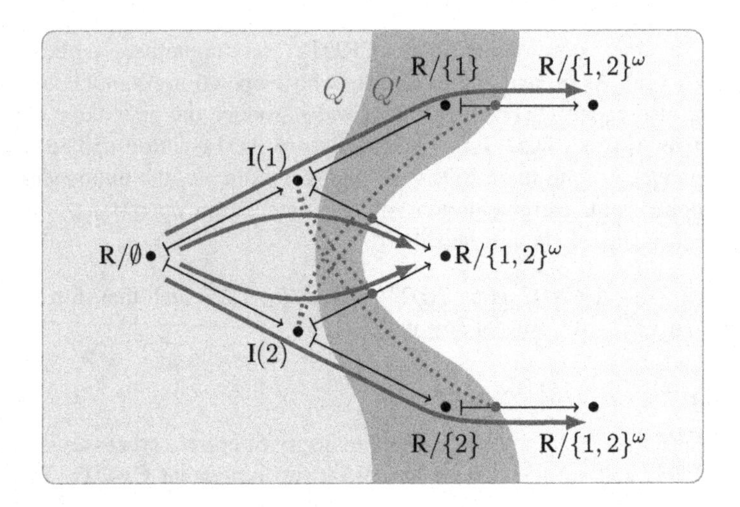

Figure 5.9. *For CADTs pipelined consistency implies eventual consistency. For the color version of this figure, see www.iste.co.uk/perrin/distributed.zip*

As H verifies pipelined consistency, for any maximal chain $p \in \mathscr{P}_H$, there exists $l_p \in \text{lin}(H[E_H/p]) \cap L(T)$. As the number of updates is finite, there is a finite prefix for l_p which contains all of these. Let Q_p be the set of queries for l_p contained in this prefix and $Q = \bigcup_{p \in \mathscr{P}_H} Q_p$.

We will now show that Q is finite. All the chains contained in Q are contained in a particular Q_p and are, thus, finite. Q may, therefore, be expressed in the form $Q = \cup_{e \in Q'} \lfloor e \rfloor_{\mapsto}$, where Q' is the set of maximal elements of Q. As $\mapsto$ is a well order, Q', which is an antichain of $(E_H, \mapsto)$, is finite and for any $e \in Q'$, $\lfloor e \rfloor_{\mapsto}$ is finite. Q, being a finite union of finite sets, is finite.

As T is a CADT, all the paths labeled by updates for H lead to the same state, ζ. Any event in the pure query $e \in E_H \setminus Q$ appearing in a linearization l_p for a given p is carried out in the state ζ, thus $H \in EC(T)$. □

5.5. Which criterion should be used?

We now know that there is no strongest weak consistency criteria and the three secondary criteria are not comparable. But does any one of them, without being stronger than the others, offer more interesting guarantees for a given program than the others? In fact, each secondary criterion has its unique advantages as well as its unique weaknesses. This is why it is important to choose the criterion that is best adapted to an application.

Serializability. In the example of instant messaging services, serializability corresponds to Hangouts (p. xiii), where Bob's message, which was not transmitted, may be qualified as "aborted". The main problem with serializability is that it does not guarantee progress: an implementation which may lead to the aborting of all the update operations and return to the initial state for all the pure query operations is serializable. This is, in fact, a relatively week criterion. To set against this, aborted events are known by the process which initiated them. This means that serializability leaves it up to the user to manage errors, by only pointing out operations that did not succeed.

In the example of instant messaging service, we understand that serializability is the most used criterion on the Internet. Services that use the client–server model naturally guarantee total order on all operations (the total order of the server). On the other hand, the operations called when not connected are lost, but the calling process is warned of this loss. The verification of the connection is generally carried out using a timer[2]. The costs of serializability are, therefore, quite high in practice, both in terms of execution time (it is necessary to wait for a response from the server) and in terms of fault tolerance (the service is a single point of failure). The existence of more efficient algorithms without servers remains an open problem.

Finally, serializability is most advisable when the application requires great control over data, for example, to carry out a bank transaction, and when the system offers enough stability to guarantee quite a low number of aborted events. This corresponds well with the client–server model, where the server is assumed to be reliable enough.

Update consistency. Both Skype (p. xv) and Facebook Messenger (p. 54) demonstrate a behavior admitted by update consistency. In both these examples, messages are reordered, whether upon receiving a new message (Skype), or later, in an asynchronous manner (Facebook Messenger). In this case, the absence of state locality is not very important as all the messages remain visible. In general, update consistency can authorize that successive reads appear completely inconsistent with respect to one another. In particular, within a distributed algorithm, written values usually depend on previously read values, which makes it difficult to use update consistency in these cases.

The impossibility of adding state locality to update consistency further impacts the cost of update consistency. In fact, no algorithm implementing update consistency can guarantee state locality. However, as its name indicates, this criterion is an abstraction of the fact that each process maintains a local state that evolves depending on the updates, in accordance with the sequential specification. These results, therefore, condemn all implementations of update consistency to using more complicated

2 Waiting for the timer does not contradict our definition of "wait-free" on the basis of fault tolerance, but we see here that some subtle changes in the definition can lead to different systems.

techniques to maintain the state in which queries are made. This explains the need to maintain a list of received messages, or to send additional correction messages in the algorithms in Chapter 3.

Finally, the criteria in the update consistency family are, above all, adapted for distributed applications themselves, as collaborative editors. In this case, the user of the shared objects is a human being who is more capable than an algorithm of adapting to small inconsistencies.

Pipelined consistency. Pipelined consistency corresponds to the strategy shown in the WhatsApp experiment (p. xiv). The problem here was that Alice and Bob did not see messages in the same order in the end, which corresponds to an absence of eventual consistency. This absence is redhibitory for many applications, especially those that manage geo-replicated data, in which maintaining a consistent state between different replication centers is a major issue.

On the other hand, the criteria in this family have much lower implementation costs than those of the other two families, as it is sufficient to broadcast a message on each update and because the local state at each process is only an abstract state (Figure 4.14).

Finally, pipelined consistency is most appropriate in cases where the processes use values that they read to decide the next values to be written. For example, in applications mostly centered on computation, as well as for the implementation of distributed algorithms such as the access to a critical section, where the necessary termination makes eventual consistency (in the long run) less important.

5.6. Conclusion

In this chapter, we have studied the structure of weak consistency criteria space. A weak consistency criterion is a consistency criterion for which any abstract data type may be implemented in the system $AS_n[\emptyset]$, made up of n asynchronous processes communicating through message passing and subject to an arbitrary number of crashes.

We have shown that there was no largest element among weak criteria. The weak consistency criteria space may be represented as a set of visible colors (Figure 5.1(a)). Like primary and secondary colors, three families of secondary criteria have, each, one family of complementary primary criteria. The conjunction of a primary criterion and its complementary secondary criterion is a strong criterion. Even if there is no unicity in this decomposition, the proposed decomposition seems particularly interesting. First of all, each primary criterion corresponds to a Consensus property. Second, the three strategies observed in the experiment in the introduction, on different instant messaging services, each reflects a family of secondary criteria. Finally, all the weak criteria studied in this work are naturally arranged into one of these families.

	PC, UC or *Ser*	EC + SL + V	EC + PC	SC
$\mathcal{W}_{k,(k \geq n)}$	$AS_n[\emptyset]$	$AS_n[t=0]$	$AS_n[t=0]$	$AS_n[t=0]$
$\mathcal{M}_{\{x,y,z\}}$	$AS_n[\emptyset]$	$AS_n[\emptyset]$	$AS_n\left[t<\frac{n}{2}\right]$	$AS_n\left[t<\frac{n}{2}\right]$
$\mathcal{M}_{\{x,y\}}$	$AS_n[\emptyset]$	$AS_n[\emptyset]$	?	$AS_n\left[t<\frac{n}{2}\right]$
CADT	$AS_n[\emptyset]$	$AS_n[\emptyset]$	$AS_n[\emptyset]$	$AS_n\left[t<\frac{n}{2}\right]$
$\mathcal{M}_x, \mathcal{W}_1$	$AS_n[\emptyset]$	$AS_n[\emptyset]$	$AS_n[\emptyset]$	$AS_n[\emptyset]$

Figure 5.10. *Minimal system necessary to implement different objects*

We also asked what the abstract data types are for which it is possible to implement a strong criterion. Figure 5.10 gives an overview of the results obtained. The objects which may be implemented wait-free are represented in blue. By varying the consistency criteria, we can highlight a hierarchy in the abstract data types. For a given system, the stronger the required consistency criterion is, the fewer the objects that may be implemented by verifying this criterion. For sliding window registers with a size of at least 2, the conjunction of the three weak criteria is impossible to implement in an asynchronous message-passing system where at least one crash can occur. Implementing the conjunction of eventual consistency and of pipelined consistency for a memory made up of at least three registers requires the hypothesis of a majority of correct processes in each execution. On the other hand, it is possible to implement a unique sequentially consistent register in wait-free systems. Whether the conjunction of eventual consistency and pipelined consistency may be implemented for a memory made up of two registers remains an open question. Finally, we have proved that the conjunction of eventual consistency and pipelined consistency may be implemented for commutative abstract data types in wait-free systems. It will be quite easy to prove that sequential consistency cannot be attained for these objects in such a system. All these results indicate that the weak criteria theory offers a new perspective from which to study and understand computability in distributed systems.

6

CODS Library

6.1. Introduction

As we mentioned in Chapter 1, a good specification technique is similar to a programming language. Identifying the abstractions that are best adapted for solving a given problem is, therefore, an important driving force in technological development. In the field of information technology, the introduction of more and more high-level programming languages has made it possible to greatly simplify programming and, as a result, greatly widen the range of applications that are produced. The invention of specialized languages such as HTML, PHP and JavaScript allowed for the emergence and the development of the Web. Developments in decentralized distributed systems have focused, above all, on strong consistency in the form of abstractions such as mutual exclusions, Consensus and even transactional memory. Abstractions still lack for weak consistency, and thus the considerable efforts needed, even today, to develop the smallest applications in systems which cannot claim to be foolproof such as peer-to-peer systems or cloud computing. We then come up against the natural question:

Problem. *What form can be taken by a language dedicated to wait-free systems?*

The approach we have followed up to now has seemed an apt choice to fulfill this role as it replaces the functional aspect at the heart of the specification of the shared objects, thanks to sequential specification. The chief strength of this approach lies in the fact that it is based on the well-established concepts of *transition systems*, *automata* and *languages* to describe sequential specifications. These mathematical concepts are central to many fields of information technology and many tools have been developed with a view to their specification, modeling and verification. Replacing sequential specification at the heart of the specification of shared objects reaffirms that all these tools may be adapted to distributed systems, at least to describe and verify the functional aspect of these programs. For instance, all the criteria based on the existence of a correct linearization vis-à-vis sequential specification (all of which reinforce validity or state locality) guarantee that all the states that have been read are accessible in the transitions system. In this paradigm, specification of sequential *objects*, in the form of *classes* or *structures*, is only natural.

Approach. *We propose reusing, without modification, all aspects of the object-oriented programming paradigm relative to sequential specifications and of only managing concurrence through the choice of consistency criterion.*

To illustrate the relevance of this approach, we present a library of consistency criteria, CODS (for "concurrent objects in distributed systems") [RUA 14]. A consistency criterion in CODS may be seen as a functor that transforms a class describing a sequential specification into another class that represents accessors to shared objects that verify this consistency criterion. Programmers can, thus, devote themselves to the description of the functional architecture of the objects in their program without worrying about the distribution aspect. They can then allow CODS to manage concurrence. In addition, CODS offers a framework that makes it possible to define new consistency criteria. This approach offers multiple advantages:

1) Using CODS greatly simplifies the development of distributed applications: a sequential object is much simpler to design than a shared object. In fact, the programmer working on sequential specification can focus exclusively on the functional aspect and set aside the problems related to concurrence.

2) The readability and evolvability of the code are much improved. If, for instance, the same abstract data type can be used in several contexts (a shared set, for example, to contain information that is useful to all processes, and a non-shared set used in local calculations within a function), the same class may be used in all contexts with only the consistency criterion changing.

3) The approach promises greater reliability in the long run: as distributed programming is more subject to bugs than sequential programming, it would seem a wise choice to entrust concurrence management to an external library that will, potentially, be better verified than a program that manages the concurrence itself.

4) Finally, we have the possibility of long-term gains in execution efficiency. It is the same as with compilers: a compiled code is, today, often better optimized than a hand-written assembly code. We can hope, similarly, that the applications that use this approach will be able to benefit more easily from progress made in generic algorithms to implement consistency criteria.

The library presented here is a prototype developed for D language [COM 16], a language from the C family whose syntax is very close to that of Java and C#.

This language was chosen with the prospect of offering the most transparent interface possible to relieve the programmer, as far as possible, of the necessity of managing the distributed aspect. D language has the three following characteristics, which are necessary for the development of such a project:

1) D is an object-oriented programming language, a prerequisite for adapting work on shared objects in a consistent manner.

2) D language offers far-reaching techniques of reflexivity, allowing for class introspection in a manner comparable to Java. This functionality is required to study the method of the sequential specification class.

3) Metaprogramming is also one of D's strong points. This functionality is required to create shared object classes at compilation time, and the methods it uses are those of the sequential specification class, which is not known *a priori*.

This chapter is presented as a tutorial for running the CODS library and may be divided into the resolution of two main challenges. On the one hand, it captures the method calls of the objects to be shared, which allows for the elegant integration of consistency criteria into an object-oriented language such as D. Section 6.2 presents the use of the library from the point of view of the programmer of a distributed application, who wishes to use consistency criteria. On the other hand, CODS offers a programming framework that allows for the definition of new criteria. Section 6.3 demonstrates how these new consistency criteria may be defined. The code for the library is available online [MOS 15].

6.2. Overview

The CODS library was designed so as to make its use as transparent as possible. In fact, other than the network configuration, which is not managed by the library, the difference between a simple sequential application and its corresponding distributed system application based on CODS essentially rests on a single line: the constructor call in the sequential program is replaced by a connection to the consistency criterion in the distributed program.

The implementation of shared objects via CODS follows the same schema as their formal specification. It is divided into a sequential implementation (a D class) and a choice of consistency criterion (from among those proposed by the library). The consistency criterion call, formatted by the sequential specification, creates a shared object that presents exactly the same interface as the sequential object.

The basic use to which the library is put is illustrated by the example in Figure 6.1, which shows a program in D that uses CODS. After including external libraries, the program code may be divided into three sections: the class `WindowRegister` that encodes a sequential specification very close to $\mathcal{W}_2$, the function p that describes the execution of the processes that use a shared object and the principal function main that launches the program.

```
1  /****************** External Libraries ******************/
2  // Standard library for D
3  import std.stdio, std.conv, core.thread;
4  // General CODS tools
5  import cods;
6  // Network simulator
7  import networkSimulator;
8  // Update Consistency
9  import uc;
10
11 /**************** Sequential Specification ****************/
12 class WindowRegister {
13   private int x = 0, y = 0;                // State
14   public void write(int val) {             // Pure update
15     x=y; y=val;
16   }
17   public string read() {                   // Pure query
18     return "<"~to!string(x)~","~to!string(y)~">";
19   }                                        // Return <x,y>
20 }
21
22 /************** Code executed by the process **************/
23 void p(int i) { // i=1 or i=2
24   // Initialisation of a shared instance
25   WindowRegister wr = UC.connect!WindowRegister("wr1");
26   // Using a shared instance
27   wr.write(i);                             // Write
28   writeln(wr.read());                      // First read
29   //Waiting for convergence
30   Thread.sleep(dur!("msecs")(1000));
31   writeln(wr.read());                      // Second read
32 }
33
34 /************ Launching parallel processes ************/
35 void main () {
36   // Declaration of the processes
37   auto network = new NetworkSimulator!2([{ p(1); }, { p(2); }]);
38   // Configuration of CODS
39   Network.configure(network);
40   // Beginning of the processes
41   network.start();
42 }
```

Figure 6.1. *Distributed program using a shared object $\mathcal{W}_2$ in D*[1]

1 In D, ~ designates the concatenation of strings of characters and ! designates a template call (line 18).

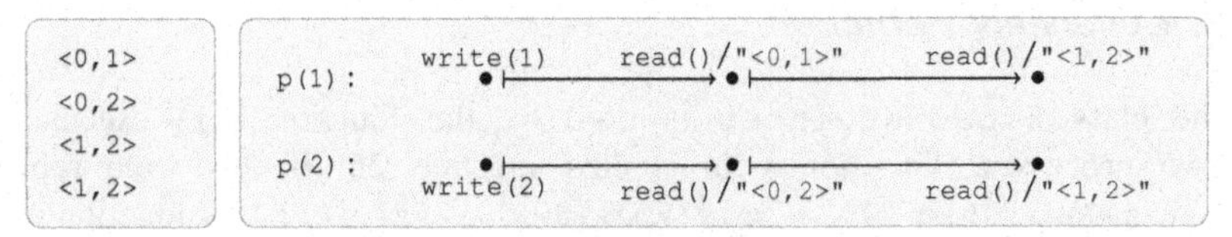

(a) Terminal. (b) Analysis of execution in the form of concurrent history.

Figure 6.2. *Output of the program given in Figure 6.1*

6.2.1. *Initialization of the network*

In order to function, CODS requires an underlying network that is capable of broadcasting messages in a reliable manner. The library itself does not manage this aspect of concurrent programming. All that it supplies is an `INetwork` interface that requires the implementation of two functions:

– `void broadcast(immutable(void)[] file)` ensures the reliable broadcast of the message `file` to all the network processes;

– `int getID()` supplies a unique identifier to the caller process.

Line 39 must be present throughout the program that uses CODS. It is required for the initialization of the network by the program. In this example, we use the class `NetworkSimulator`, which simulates an asynchronous network using UNIX processes. Lines 37 and 41 are relative to this class. The only important information that concerns them is that they launch two parallel processes. One executes the function `p(1)` and the other executes the function `p(2)`.

6.2.2. *Sequential specification*

These two processes use an object of the type `WindowRegister` whose class is defined between lines 12 and 20. Such a class encodes an ADT. The set of states is the space for the definition of the member variables, abstracted here as $\mathbb{N}^2$. The initial state is the default value of the member variables, in this case $(0, 0)$. The input alphabet A contains the member methods with all the possible combinations for their arguments. We thus find the symbols `read()` and `write(n)` for every $n \in \mathbb{N}$ within A. The output alphabet, B, is the union of the possible return values for all the methods, the type `void` being replaced by a unitary type that contains a single value $\perp$. In this example, B contains $\perp$ and all the character strings of the form "`<x,y>`", where $x, y \in \mathbb{N}$. Finally, the transition function and the output function are defined by the side-effect and return value of the class methods. The ADT specified by the class `WindowRegister` is, therefore, nearly identical to $\mathcal{W}_2$ (see p. 7), to the name of operations and to the query return type.

6.2.3. *Consistency criterion*

The class `WindowRegister` is used by the function p(i) executed by the two processes. The library is applied on line 25. If this were replaced by `WindowRegister wr = new WindowRegister();` this function would be nothing special. It would begin with the creation of an object of the type `WindowRegister`, would write its argument in it and would then read twice, with a one-second interval. The only unique feature of CODS here is the manner in which the shared objects are initialized. The constructor call is replaced by the instruction `UC.connect!WindowRegister("wr1")`, which contains three pieces of information:

– `UC` is the consistency criterion used. This refers to update consistency. In order to be able to use this criterion, we need to import the corresponding library, which is done in line 9. For the time being, only update consistency and pipelined consistency are implemented;

– `WindowRegister` is the class that corresponds to sequential specification;

– "`wr1`" is the unique identifier of the shared object on the network. In this example, the two processes initialized their instance with the same identifier and their operations, consequently, are concerned with the same shared object. If a process connects itself several times with the same unique identifier, it will obtain the same object.

One way of interpreting this instruction is to imagine that all the objects pre-exist virtually in the network. The instruction only creates an accessor that allows them to be manipulated, of the type `UC.Type!WindowRegister`, inherited from `WindowRegister`. This allows for the perfect integration of the object with its environment, as there is no other modification required in the rest of the program.

The program is not deterministic. The output given in Figure 6.2(a) is one of the possible outputs. What happened during this execution may be explained by a concurrent history (Figure 6.2(b)). The process that executes `p(1)` wrote 1 in the shared object and then read "`<0,1>`" before convergence and "`<1,2>`" after convergence (a wait-time of one second was enough to attain convergence in all the cases tests). The process that executed `p(2)` wrote 2, then read "`<0,2>`" and "`<1,2>`". There was, thus, an exchange of information between the processes and the history obtained verifies update consistency but not sequential consistency. Update consistency is guaranteed for all executions that use the consistency criterion `UC`.

6.2.4. *Composition of objects*

As we saw in Chapter 1, composability is a rare property for consistency criteria. It would be too restrictive to only accept composable criteria. Instead of applying the consistency criterion directly to each sequential specification in the hope that they behave mutually as if the criterion applied to the composition of the specifications, CODS composes the sequential specifications before applying the consistency criterion. In other words, if a program uses two ADTS, T_1 and T_2, with a consistency criterion C, CODS will implement $C(T_1 \times T_2)$ rather than $C(T_1, T_2)$. We will designate the composition of the sequential specifications of all objects used in the program by ΠT.

When the p function of the program in Figure 6.1 is replaced by the function in Figure 6.3, the two processes manage two shared objects wr_1 and wr_2, which have the identifiers "wr1" and "wr2", respectively. The process that executes $p(1)$ writes 2 in wr_1 then 3 in wr_2, and the other process writes 4 in wr_2 then 5 in wr_1. After convergence, it is impossible to simultaneously have wr_1 in the state $(5,2)$ and wr_2 in the state $(3,4)$, as the history obtained would be in $UC(wr_1, wr_2)$ but not in $UC(wr_1 \times wr_2)$.

```
1  void p (int i) {
2    WindowRegister wr1 = UC.connect!WindowRegister("wr1");
3    WindowRegister wr2 = UC.connect!WindowRegister("wr2");
4    if(i==1) {                                  // code for p(1)
5      wr1.write(2);
6      wr2.write(3);
7    } else {                                    // code for p(2)
8      wr2.write(4);
9      wr1.write(5);
10   }
11   // Waiting for convergence
12   Thread.sleep(dur!("msecs")(500));
13   // <5,2> and <3,4> are forbidden convergence
         states
14   writeln(wr1.read() ~ " " ~ wr2.read());
15 }
```

Figure 6.3. *Program using a single object made up of two sliding window registers*

6.2.5. *Transactions*

The composition of the sequential specifications implemented in CODS makes it possible to correctly manage composition without impacting modularity, as it is transparent. Composition also allows for an advanced CODS functionality: *transactions*. A transaction in CODS is a piece of code that makes it possible to encapsulate a succession of operations on shared objects in a unique concurrent history event. From the point of view of the consistency criterion, a transaction behaves exactly like an operation. The only difference is that it is not defined directly in the class of shared objects, but is generally defined within the code that governs process behavior. In CODS, the set of transactions defined in the program, Γ, applies to the composition of the sequential specifications of shared objects, ΠT, which creates a new ADT $\Pi T[\Gamma]$, in which all the transactions have been transformed into operations. The formalization of the concept of transaction and of their interaction with shared objects is shown in definition 6.1.

DEFINITION 6.1.– Let there be an ADT $T = (\mathsf{A}, \mathsf{B}, \mathsf{Z}, \zeta_0, \tau, \delta) \in \mathcal{T}$. A *transaction* on T is a triplet $\gamma = (\mathsf{B}_\gamma, \tau_\gamma, \delta_\gamma)$, where:

– B_γ is a countable set called *output alphabet*;

– $\tau_\gamma : \mathsf{Z} \rightarrow \mathsf{Z}$ is a transitions function;

– $\delta_\gamma : \mathsf{Z} \rightarrow \mathsf{B}_\gamma$ is the output function.

Let Γ be a countable set of transactions on T. The application of Γ on T is the ADT $T[\Gamma] = (\mathsf{A}' = \mathsf{A} \cup \Gamma, \mathsf{B}' = \mathsf{B} \cup \bigcup_{\gamma \in \Gamma} \mathsf{B}_\gamma, \mathsf{Z}, \zeta_0, \tau', \delta')$ where:

$$\tau' : \begin{cases} \mathsf{Z} \times \mathsf{A}' \rightarrow & \mathsf{Z} \\ (q, \alpha) \mapsto \tau(q, \alpha) \text{ if } \alpha \in \mathsf{A} \\ (q, \gamma) \mapsto \tau_\gamma(q) \text{ if } \gamma \in \Gamma \end{cases} \delta' : \begin{cases} \mathsf{Z} \times \mathsf{A}' \rightarrow & \mathsf{B}' \\ (q, \alpha) \mapsto \delta(q, \alpha) \text{ if } \alpha \in \mathsf{A}. \\ (q, \gamma) \mapsto \delta_\gamma(q) \text{ if } \gamma \in \Gamma \end{cases}$$

CODS offers two transaction types: *named transactions*, the most general, and *anonymous transactions*, which are simpler to use although more restricted.

6.2.6. *Named transactions*

Named transactions are a mechanism that makes it possible to define all the transactions that apply on shared objects. A named transaction is a class that inherits the abstract class `Transaction!T`, where `T` is the return type for the transaction (B_γ). It has to implement the method `T execute()`, which defines the behavior of the transaction.

```
1  /****************** External libraries *****************/
2  import std.stdio, std.conv, core.thread;
3  import cods, networkSimulator, uc;
4
5  /************* Sequential specification ***************/
6  class Register(T) {
7    private T state = 0;
8    public T write(T val) {           // Pure update
9      state=val; return val;
10   }
11   public T read() {                 // Pure query
12     return state;
13   }
14 }
15
16 /********** Specification of the named transaction **********/
17 class Increment : Transaction!void {   // void : return type
18   private int v;
19   public this(int val) {               // Constructor
20     v = val;
21   }
22   public override void execute() {     // Obligatory
23     Register!int x = UC.connect!(Register!int)("RegX");
24     x.write(v + x.read()); // int tmp=x.read(); x.write(v+tmp);
25   }
26 }
27
28 /************** Code executed by the processes *************/
29 void p(int i) {
30   Register!int x = UC.connect!(Register!int)("RegX");
31   // Using the transaction
32   UC.transaction!void(new Increment(i));
33   Thread.sleep(dur!("msecs")(500));
34   writeln(x.read());
35 }
36
37 /************ Launching the parallel processes ***********/
38 void main () {
39   Network.registerType!Increment; // Obligatory declaration
40   auto network = new NetworkSimulator!2([{p(1);}, {p(2);}]);
41   Network.configure(network);
42   network.start();
43 }
```

Figure 6.4. *Implementation of incrementation using a named transaction*

The program given in Figure 6.4 illustrates the use of named transactions. In addition to the inclusion of libraries, the definition of sequential specification `Register(T)` and the principal function of the program, which are similar to the program given in Figure 6.1, the program contains the definition and call for the transaction `Increment`:

– the named transaction `Increment` is defined between lines 17 and 26. This transaction makes it possible to increase a register by a value `val` that is passed as an argument to the constructor. The body of the function `execute()` contains the operations that are required for the incrementation of a register. The connection to this register is carried out in the usual manner in line 23;

– the `C.transaction!T(Transaction!T t)` function is used in line 32 to call the transaction. `C` is the consistency criterion and `T` is the return type for the transaction `t` (`void` in the case of the transaction `Increment`). Let us note that it is necessary to declare the transaction type at the beginning of the program. In the example, this is done in line 39.

In this program, one process increases the register by 1 and another increases it by 2. If the call for the transaction were replaced by `x.write(i + x.read())`, the read and write for the two processes could be interleaved such that the register converges towards 1, 2 or 3. By encapsulating the calculation inside a transaction, we guarantee that the convergence value must be 3.

```
1  void p(int i) {
2    WindowRegister x = UC.connect!WindowRegister("x");
3    UC.anonymousTransaction({       // Start of transaction
4      x.write(i);
5      x.write(i);
6    });                             // End of transaction
7    Thread.sleep(dur!("msecs")(500));
8    writeln(x.read());              // "<1,1>" or "<2,2>" possible
9  }
```

Figure 6.5. *Example of an anonymous transaction*

6.2.7. *Anonymous transactions*

Defining a new class for each transaction may sometimes excessively complicate the source code for a program. *Anonymous transactions* are a light mechanism that allow for the definition of transactions in the form of a *succession of independent pure update operations*.

An anonymous transaction may be defined at the same spot where it is called. The function `C.anonymousTransaction(void delegate() dg)` must be called to create an anonymous transaction, with `C` being the desired consistency criterion. In D, the type `RT delegate(T)` designates an anonymous function whose return type is `RT` and whose argument is of the type `T`. The object `dg` is an anonymous function whose body, defined within brackets, contains the sequence of operation calls that compose the transaction. In the program given in Figure 6.5, there can be no interleaving of operations of the two processes. The final state, thus, can only be $(1,1)$ or $(2,2)$ while without the transactions we could have converged towards $(1,2)$ or $(2,1)$.

Anonymous transactions are much simpler to use than named transactions, but their usage is also governed by more restrictions. This is due to restrictions on the serialization library for D language, Orange [CAR 14], which we use to create messages, and which is not applicable to objects of the type `delegate`. An anonymous transaction is mandatorily formed of a succession of independent operations on shared objects. In particular, the value returned by one operation cannot be used as an argument in the following operation. For example, the transaction in Figure 6.4 cannot be written as an anonymous function as the value written depends on the value read. Anonymous transactions are, nonetheless, of interest, chiefly to prepare a state by preventing certain race conditions. If, for example, a program manages two shared sets, A and B, and wants to always guarantee that $A \subset B$, it must always insert the values in B when it wants to insert them in A, and delete them in A to delete them from B. It can also use anonymous transactions to avoid interleavings of the type $A.\mathrm{D}(x) \cdot A.\mathrm{I}(x) \cdot B.\mathrm{I}(x) \cdot B.\mathrm{D}(x)$, which lead to a forbidden state. It is for the same reason that a transaction cannot return values, as the instruction `return` is not an operation on a shared object. This is why anonymous transactions can only be pure updates.

6.3. Defining new criteria

CODS also defines a framework that makes it possible to implement new consistency criteria. Figures 6.6 and 6.7 demonstrate the implementation of pipelined consistency in this framework. The use of this criterion does not differ from update consistency except in the importing of external libraries (`import pc;` replaces `import uc;`) and in the criterion used during the call to CODS functionalities. (`PC.connect`, `PC.anonymousTransaction` and `PC.transaction`).

The class `PC` is defined in line 14 as an empty derived class inherited from `ConsistencyCriterionBase!PCImplementation` (i.e. an alias). The class `ConsistencyCriterionBase` provides the CODS interface. It is in charge of capturing the method calls for shared objects, of encapsulating these calls and the transactions in a common form and of transmitting them upon the implementation of the shared criteria. It also manages local concurrence within the process to guarantee the atomic execution of the functions in the implementation of the criterion. The main job to be provided to implement the criterion is the class `PCImplementation`, defined between lines 16 and 54.

```
1  import cods;
2  /************ Message to broadcast on the network ************/
3  class PC_Message : Message {
4    private int id, cl; private Operation op; // Data
5    this(int mId, int mCl, Operation mOp) { // Constructor
6      id=mId; cl=mCl; op=mOp;
7    }
8    override void on_receive() {           // Called upon reception
9      PC.getInstance().getImplementation()
10                              .receiveMessage(id, cl, op);
11   }
12 }
13 /************* Defining PC as an abbreviation ***************/
14 class PC : ConsistencyCriterionBase!PC_Implementation {};
```

Figure 6.6. *Implementation of pipelined consistency: file* `pc.d` *(beginning)*

This contains the algorithm, which implements the consistency criterion. It must implement the nested class `SharedObject!T` and the method `executeOperation()` to extend the abstract class `ConsistencyCriterionImplementation`. When a method or a transaction is executed at the level of the user program, the method `executeOperation` is called locally with a parameter that encodes the complex operation. The operation may be executed (locally or otherwise) thanks to its `execute()` methods (lines 27 and 41). This execution generates the succession of methods to be called for the objects. Thus, the `SharedObject!T.executeMethod()` method is called for the corresponding object. For instance, for each execution of the transaction defined in the program in Figure 6.4, the executeOperation method is called once by the process that has ordered the transaction and the `SharedObject!Register.executeMethod()` method is called twice for each process: once for the read and once for the write. In the case of pipelined consistency, the `SharedObject!T` class is very simple. It encodes the local state of a shared object with an abstract state of its ADT (line 48) and restricts itself to applying methods to this local state when it receives them (line 51). Let us note that the return type, `ExtObject`, makes it possible to contain all the types, including `void` and the basic types.

The class `PCImplementation` is chiefly an implementation of the FIFO broadcast based on reliable broadcast. When an operation is carried out, a `PCMessage(mId, mCl, mOp)` message is broadcast immediately following its local execution. In order to be delivered, a message must extend the abstract class `Message` and yield an `onreceive()` method. Upon receiving this message, the `receiveMessage` method is called. The implementation of FIFO reception is standard and is based on a vector clock, `clock`, that indicates the number of messages received from each process and a buffer, `pending`, which stocks messages that were received but whose corresponding writes have not yet been applied. We can see that the message class must also be recorded (line 20) so that it can be decoded.

```
15 /********* Principal class: implementation of PC *********/
16 class PC_Implementation : ConsistencyCriterionImplementation {
17   private int[int] clock;                    // Vector clock
18   private Operation[int][int] pending;       // Message buffer
19   this() {                                   // Constructor
20     Network.registerType!PC_Message;         // Registration of type
21     clock = [Network.getInstance().getID() : 0];
22   }
23   // An operation or a transaction was locally called
24   ExtObject executeOperation(Operation op) {
25     int id = Network.getInstance().getID();
26     Message m = new PC_Message(id, clock[id]+1, op);
27     ExtObject o = op.execute();              // Atomic local execution
28     clock[id] = clock[id]+1;
29     Network.getInstance().broadcast(m);      // Broadcast of message
30     return o;
31   }
32   // A PC_Message(mId, mCl, mOp) message was received
33   void receiveMessage(int mId, int mCl, Operation mOp) {
34     if(!(mId in clock)) clock[mId] = 0;      // First message of mId
35     if (clock[mId] < mCl) {                  // If not already received
36       pending[mId][mCl] = mOp;               // Adding a message to the buffer
37       // while there remain messages in the buffer
38       for(int cl = clock[mId] + 1; cl in pending[mId]; cl++){
39         clock[mId] = cl;                     // Delivery of message
40         pending[mId].remove(cl);             // Clean-up of the buffer
41         pending[mId][cl].execute();          // Local execution
42       }
43     }
44   }
45   // Internal representation of shared objects
46   override static public class SharedObject(T) :
47       ConsistencyCriterionImplementation.SharedObject!T {
48     T t = new T();                           // Local state of an instance
49     // Function called for each method inside the transaction
50     override public ExtObject executeMethod(Functor!T f) {
51       return f.execute(t);                   // Local execution
52     }
53   }
54 }
```

Figure 6.7. *Implementation of pipelined consistency: file `pc.d` (end)*

6.4. Conclusion

In this chapter, we have introduced the CODS library, which shows how the theory behind weak consistency criteria can be put into programming languages such as D. The library is designed keeping in mind two objectives:

– it must have the simplest possible interface: in the simplest programs, only the instantiation of objects is affected by the library;

– the library is created so as to be extendable: a framework makes it possible to implement new criteria and the network configuration continues to be managed by the developer.

Unfortunately, this adaptability risks becoming the first limitation of this approach. The most efficient algorithm to implement a criterion depends on the system on which the program is executed. For instance, in a synchronous, parallel system, we can know at all times that a message that is older than a certain date cannot be received. This greatly simplifies the implementation of update consistency. Additionally, it is currently impossible to adapt an algorithm based on the objects to be implemented. We also know that some objects, such as sets and memory, have particular structures that make large optimizations possible. Pure queries and pure updates must also be processed differently. One possible solution is to add annotations such as `@update` and `@query` to declare operation properties in sequential specifications. To go much further and have as much scope for action as a compiler, it seems essential to integrate consistency criteria within the compiler itself or design a dedicated language. This would also make it possible to revisit disagreements that are linked to technical difficulties related to D language rather than design choices, for instance, the necessity of recording transaction types and message types and the restrictions imposed by anonymous transactions.

Moreover, CODS composition policy is interesting as it allows for the use of non-composable consistency criteria in a straightforward manner. This strategy may, however, be restrictive with respect to non-decomposable criteria. In this case, some objects taken separately may not verify the consistency criterion. It is up to the developer of the consistency criterion to ensure that the implementation chosen prevents this situation from arising. In the case of pipelined consistency, this problem does not arise, as the criterion is decomposable. However, update consistency is not decomposable and the algorithm $UC[k]$ does not guarantee that the object of each composition will verify update consistency. The quest for an efficient and decomposable algorithm to implement update consistency is still ongoing.

Conclusion

Summary

Shared objects are essential to distributed systems because they are able to model all levels of any application, from the communication primitives to the application in its entirety. In asynchronous wait-free message-passing systems, it is impossible to implement strong consistency criteria such as sequential consistency or linearizability: it must be accepted that it is inevitable that some inconsistencies will arise. This makes the specification of these objects and their inconsistencies more important as well as more complicated. And thus the question we have closely studied in this book has been: *how can we specify shared objects in a wait-free system?*

We suggest retaining the same schema for the specification of these shared objects, separating sequential specification into two complimentary facets: a sequential specification and a consistency criterion. Sequential specification is chosen as a basis for this because it is possible to reuse well-established concepts from automata and formal languages to study them, and also because they correspond to usual concepts in a programming language. This last point is illustrated by the CODS library for D language, according to which any class may play the role of sequential specification. The consistency criterion to be applied to the class is only chosen from among those offered by the library. The use of CODS is almost transparent: only the instantiation of objects is replaced by a connection to a pre-existing shared instance, through a consistency criterion.

Choosing the best criterion to apply in a given situation remains a problem that requires know-how. Our study allows us to group existing criteria into several large families. A consistency criterion that is weaker than sequential consistency and for which it is impossible to implement all the objects in a wait-free system is called a *weak criterion.* We have studied weak criteria space as a mathematical object. In the form of a truncated cone, it possesses no single largest element. We studied three families of weak criteria, called *secondary criteria*, which, when conjugated pairwise, result in a strong criterion. Each secondary criterion has a complementary *primary* criterion that makes explicit the type of inconsistency that it will admit. This study is centered on wait-free systems, but the main proof given in Chapter 5 is based solely on the impossibility of Consensus, and it holds true, therefore, for many other systems.

Pipelined Consistency/Eventual Consistency. In a history that verifies pipelined consistency, processes are not concerned with queries carried out by other processes. As a result, different processes may apply operations in different orders. This prevents eventual consistency. Pipelined consistency is especially useful for objects used by parallel programs, such as scientific simulations: each query is important, being used in calculations in the rest of the program. Eventual consistency is of low importance, either because executions are supposed to terminate or because the process never stops updating. A typical example for their use is the synchronization between processes, for example, to control the accesses to a critical section.

Update consistency/State locality. Update consistency reinforces eventual consistency by forcing a convergence state to be one of the states admissible by sequential consistency. This criterion as well as its strong variant were studied in detail in Chapter 3, both being implementable in wait-free systems. To guarantee convergence, update consistency is sometimes forced to revisit its choices, which may lead it to contradict the rules set by sequential specification (violating state locality). It is thus more appropriate for collaborative applications, as human users can adapt to inconsistencies and correct them. However, they will require eventual consistency for agreement on a final document.

Serializability/Validity. The operations of a serializable history may be aborted or accepted. Serializability offers strong guarantees on the accepted operations, which form a sequentially consistent history. On the other hand, there is no limit on the number of aborted operations, and therefore this does not guarantee that it will always be possible to execute operations. Additionally, in order to bring down the number of aborted operations, it may be necessary to move away from the wait-free model. To summarize, serializability is most recommended for applications where it is of prime importance to avoid inconsistencies that may even momentarily prevent a program from working.

In theory, we can construct an infinity of weak criteria that are impossible to conjugate pairwise. Beyond this purely theoretical observation, we believe that the three secondary criteria are of particular importance. On the one hand, each one corresponds to a way of weakening one of the Consensus properties, which is known to be a fundamental problem with distributed algorithms. On the other hand, we have been able to identify widely used instant messaging services whose strategies when dealing with a loss of connection illustrate all three types of inconsistency.

Each family has its own internal structure. We have had a brief glimpse into these in Chapter 4 when studying causality. This does not form a separate family, but may reinforce other criteria within their family. We have defined weak causal consistency in the validity family, causal convergence in the update consistency family and causal consistency and strong causal consistency in the pipelined consistency family. The map in Figure 1 summarizes the main criteria discussed.

Finally, the study of weak criteria is useful in order to understand computability in distributed systems. In fact, the most important hierarchy of shared objects to date, Herlihy's hierarchy, is based on the capacity of an object to resolve Consensus. We have shown that there was a way of refining the first level of the hierarchy by varying the consistency criterion. We also identified the CADT class, for which pipelined consistency is stronger than eventual consistency.

Future work

This book raises many questions that have not been explored. We will now list some of them.

Mapping weak criteria space

Weak consistency is often summarized by a hierarchy of criteria: eventual consistency is the weakest, pipelined consistency reinforces it by taking into account process order, then we have causal order, which further adds semantic relations between updates and queries, and finally, we have strong consistency. A deeper study of weak criteria space reveals quite a different picture. Can we transpose this hierarchy into each family? We saw that causality was a structuring component of several families. What about process order? Is there any intermediary criterion between validity and weak causal consistency that takes this into account? What would be the result if we composed this criterion with update consistency? Conversely, can we take it out from pipelined consistency? And if so, would the resulting criterion always be a good representative of its family, or, on the contrary, would it be possible to implement memory that verifies both eventual consistency and this new criterion? Apart from causality and FIFO order, are there any other common properties that give a similar structure to multiple families?

SC	PC	SEC	EC	CCv	WCC	SCC
Sequential Consistency (p. 28)	Pipelined Consistency (p. 46)	Strong Eventual Consistency (p. 39)	Eventual Consistency (p. 37)	Causal Convergence (p. 87)	Weak Causal Consistency (p. 84)	Strong Causal Consistency (p. 97)

$C_\perp$	SL	SUC	UC	Ser	V	CC
Minimial Criterion (p. 19)	State Locality (p. 110)	Strong Update Consistency (p. 61)	Update Consistency (p. 60)	Serializability (p. 35)	Validity (p. 109)	Causal Consistency (p. 90)

Figure 1. *Mapping weak criteria space. For the color version of this figure, see www.iste.co.uk/allard/systems.zip*

Dependency on systems

Even though one of our aims was to avoid, as far as possible, references to the systems on which algorithms were implemented, several results are linked to the system $AS_n[\emptyset]$. A point to ponder is how they would be affected by a change in system.

From a theoretical point of view, the secondary criteria families remain incompatible as long as Consensus is impossible. On the other hand, it is not a given that all families will always be implemented here. In a system that contains forking and joining threads, which translates into events with several concurrent antecedents according to the process order, it is essential that all the processes are in the same state before the join so that the history is pipelined consistent. Pipelined consistency is, therefore, as difficult to implement in these systems as its conjunction with eventual consistency.

From a practical point of view, the algorithms that are implemented differ greatly from system to system and thus their complexity also varies. For instance, with a maximal time on the delivery duration for messages, we can considerably diminish the complexity of update consistency.

Specific implementations

In this book, we chiefly focused on generic algorithms that function regardless of the prescribed sequential specification. For a set with strong update consistency, we saw that a dedicated algorithm can be much more efficient. It would be interesting to have optimized algorithms for other data types.

In the long term, we can imagine that CODS will itself be able to find the best optimizations based on the sequential specification code. For this to happen, complex semantic analysis techniques must be implemented. The introspective capabilities of D language are probably not adequate for this and it will certainly be necessary to modify a language compiler.

Quantitative evaluation of consistency

A weaker consistency criterion is always cheaper to implement than a stronger criterion as the best implementation of a strong criterion is also an implementation of all the weaker criteria. However, is the theoretical gain using weak criteria for which all operations are wait-free also practically beneficial? Wait-free implementations generally require more resources in terms of memory and even communication, which slows down the executions. It is very difficult to compare the different consistency criteria equally as they are generally meant for different systems or even different applications. However, it seems important to be able to reliably evaluate the assertions of efficiency that are made regarding weak consistency, which are often presented as obvious but are never really explained in detail.

Apart from cost comparisons, it would be interesting to quantitatively compare consistency brought about by the implementation of criteria. For example, we said on p. 109 that serializability was a weak criterion, as it was possible to get it to abort all updates. However, such an implementation has no practical use. How do you define the metrics to measure the distance of an implementation with a stronger consistency criterion, such as sequential consistency?

Resolving integrity constraints

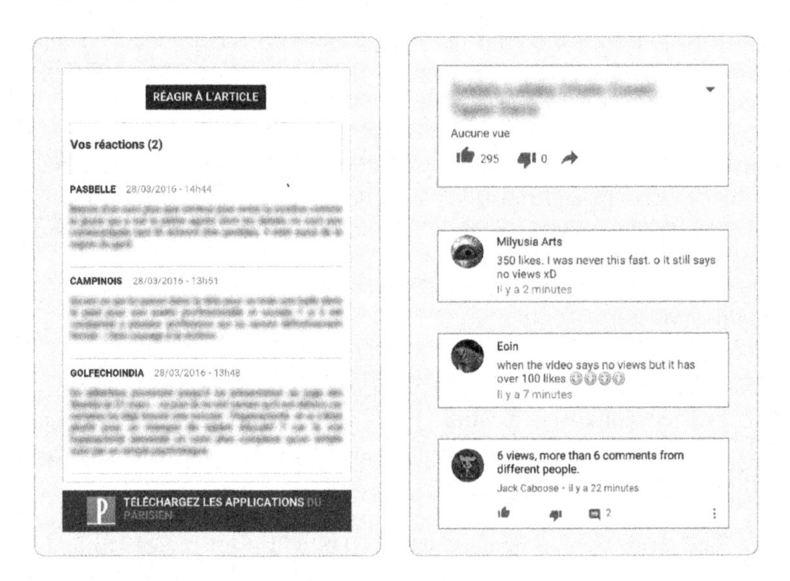

Figure 2. *Integrity violation on the sites for Le Parisien and on YouTube*

The absence of composability and the choice of consistency criteria that are too weak may lead to queries that return incomprehensible results. Figure 2 depicts two illustrations of this phenomenon on websites. On the left, we have a section of the comments below an article in a French daily, Le Parisien[2], which indicates that only 2 commentators have posted, while three comments are displayed. On the right, we have a frequent occurrence from the video-hosting site YouTube[3] that indicates a much

2 http://www.leparisien.fr/

3 https://www.youtube.com/

lower number of views than the number of votes for a video. This is strange because each member can vote only once for a video and it is almost impossible to vote for a video without having been counted as a viewer. These inconsistencies are often pointed out by visitors to the sites, as can be seen in the YouTube comments in the example[4].

For databases, such inconsistencies are grouped under the term *integrity constraints*, which may be defined as invariants for the set of accessible states of the abstract data types. Most of the consistency criteria relate queries to linearizations of updates, that is, they impose a form of accessibility on abstract data types. Damien Maussion, in an internship in the first year of his Master's program, proposed a data structure that integrates these constraints and verifies update consistency, by first using the CODS library and then developing a dedicated algorithm [MAU 15]. He could, thus, demonstrate that it would be possible to verify domain constraints, unicity constraints and referential constraints (which is enough to implement foreign keys), without synchronization.

Hybrid criteria

We saw in Chapter 2 that several consistency criteria were hybrid criteria formed by several other weak criteria. For now, we have only considered directly defined criteria, but their hybridization is an important question. For example, what would be the behavior of two objects that verify different consistency criteria within the same program? If one of the criteria is stronger than the other, we would like the composition to verify at least the weakest criterion.

We also have the question of two operations verifying different criteria. To go back to the example of instant messaging services, the deletion of a message is usually local: only the user who deletes a message sees it disappear from their message queue. If we take this new operation into account, eventual consistency can only be guaranteed if no message is deleted. How can we model this phenomenon? How can it be added to a programming interface like CODS?

Moving towards a new hierarchy of sequential objects?

We have seen that the study of weak criteria was a good tool to study computability in distributed systems. More generally, we can ask whether there is a sequential hierarchy of objects which ranks objects based on how difficult they are to

---

4 In the images for the YouTube page, the three comments and the statistics are from different screenshots (the last comment was posted on a different video). However, the page was not refreshed between the shot of the statistics and the first comment, which shows the absence of causality on the positive vote counter, which is assumed to be greater.

implement in wait-free systems and, if this hierarchy exists, whether it would be compatible with Herlihy's hierarchy. To illustrate these questions, we propose the conjecture 1.

CONJECTURE 1.– Let $T \in \mathcal{T}_{UQ}$ be an UQ-ADT such that $(PC + EC)(T)$ can be implemented in $AS_n[\emptyset]$. Then, $SC(T)$ may be implemented in $AS_n[t < \frac{n}{2}]$.

If this conjecture is true, we have already identified two categories of objects whose Consensus number is 1, defined based on consistency criteria and in very different systems. They also refine Herlihy's hierarchy. We know that this class contains all the CADTs and the single register, but not memory made up of three registers. What about two registers?

Bibliography

[ADV 90] ADVE S.V., HILL M.D., "Implementing sequential consistency in cache-based systems", *Proceedings of the 1990 International Conference on Parallel Processing*, vol. 1, pp. 47–50, 1990.

[ADV 96] ADVE S.V., GHARACHORLOO K., "Shared memory consistency models: a tutorial", *Computer*, vol. 29, no. 12, pp. 66–76, IEEE, 1996.

[ADV 10] ADVE S.V., BOEHM H.-J., "Memory models: a case for rethinking parallel languages and hardware", *Communications of the ACM*, vol. 53, no. 8, pp. 90–101, ACM, 2010.

[AGU 07] AGUILERA M.K., FROLUND S., HADZILACOS V. *et al.*, "Abortable and query-abortable objects and their efficient implementation", *Proceedings of the Twenty-Sixth Annual ACM Symposium on Principles of Distributed Computing*, ACM, pp. 23–32, 2007.

[AHA 93] AHAMAD M., BAZZI R.A., JOHN R. *et al.*, "The power of processor consistency", *Proceedings of the Fifth Annual ACM Symposium on Parallel Algorithms and Architectures*, ACM, pp. 251–260, 1993.

[AHA 95] AHAMAD M., NEIGER G., BURNS J.E. *et al.*, "Causal memory: definitions, implementation, and programming", *Distributed Computing*, vol. 9, no. 1, pp. 37–49, Springer, 1995.

[AIY 05] AIYER A., ALVISI L., BAZZI R.A., "On the availability of non-strict quorum systems", *Distributed Computing*, pp. 48–62, Springer, 2005.

[ALV 11] ALVARO P., CONWAY N., HELLERSTEIN J.M. *et al.*, "Consistency analysis in bloom: a CALM and collected approach", *Proceedings of 5th Biennial Conference on Innovative Data Systems Research*, pp. 249–260, 2011.

[ALV 12] ALVARO P., MARCZAK W.R., CONWAY N. *et al.*, "Logic and Lattices for Distributed Programming", *Proceedings of the Third ACM Symposium on Cloud Computing*, pp. 1–14, 2012.

[ASL 11] ASLAN K., MOLLI P., SKAF-MOLLI H. *et al.*, "C-set: a commutative replicated data type for semantic stores", *RED: Fourth International Workshop on Resource Discovery*, 2011.

[ATT 94] ATTIYA H., WELCH J.L., "Sequential consistency versus linearizability", *ACM Transactions on Computer Systems (TOCS)*, vol. 12, no. 2, pp. 91–122, ACM, 1994.

[ATT 95] ATTIYA H., BAR-NOY A., DOLEV D., "Sharing memory robustly in message-passing systems", *Journal of the ACM (JACM)*, vol. 42, no. 1, pp. 124–142, ACM, 1995.

[ATT 15] ATTIYA H., ELLEN F., MORRISON A., "Limitations of highly-available eventually-consistent data stores", *Proceedings of the 2015 ACM Symposium on Principles of Distributed Computing*, ACM, pp. 385–394, 2015.

[BAI 12] BAILIS P., FEKETE A., GHODSI A. *et al.*, "The potential dangers of causal consistency and an explicit solution", *Proceedings of the Third ACM Symposium on Cloud Computing*, ACM, p. 22, 2012.

[BAL 04] BALDONI R., MILANI A., PIERGIOVANNI S.T., "An optimal protocol for causally consistent distributed shared memory systems", *Parallel and Distributed Processing Symposium, 2004. Proceedings of 18th International*, IEEE, p. 68, 2004.

[BAL 06] BALDONI R., MALEK M., MILANI A. *et al.*, "Weakly-persistent causal objects in dynamic distributed systems", *25th IEEE Symposium on Reliable Distributed Systems, 2006. SRDS'06*, IEEE, pp. 165–174, 2006.

[BER 83] BERNSTEIN P.A., GOODMAN N., "Multiversion concurrency control-theory and algorithms", *ACM Transactions on Database Systems (TODS)*, vol. 8, no. 4, pp. 465–483, ACM, 1983.

[BER 91] BERSHAD B.N., ZEKAUSKAS M.J., Midway: shared memory parallel programming with entry consistency for distributed memory multiprocessors, Report no. CMU-CS-91–170, Carnegie Mellon University, 1991.

[BIE 12] BIENIUSA A., ZAWIRSKI M., PREGUIÇA N. *et al.*, An optimized conflict-free replicated set, Report, Pierre and Marie Curie University, 2012.

[BIR 87] BIRMAN K.P., JOSEPH T.A., "Reliable communication in the presence of failures", *ACM Transactions on Computer Systems (TOCS)*, vol. 5, no. 1, pp. 47–76, ACM, 1987.

[BRE 91] BREITBART Y., GEORGAKOPOULOS D., RUSINKIEWICZ M. *et al.*, "On rigorous transaction scheduling", *IEEE Transactions on Software Engineering*, vol. 17, no. 9, pp. 954–960, IEEE, 1991.

[BRE 00] BREWER E.A., "Towards robust distributed systems", *PODC*, vol. 7, 2000.

[BUR 12] BURCKHARDT S., LEIJEN D., FÄHNDRICH M. *et al.*, "Eventually consistent transactions", *Programming Languages and Systems*, Springer, pp. 67–86, 2012.

[BUR 14] BURCKHARDT S., GOTSMAN A., YANG H. *et al.*, "Replicated data types: specification, verification, optimality", *Proceedings of the 41st Annual ACM SIGPLAN-SIGACT Symposium on Principles of Programming Languages*, ACM, pp. 271–284, 2014.

[CAR 05] CARTWRIGHT J., "Gare Du Nord departures", available at: https://www.flickr.com/photos/43917222@N00/16226740, (accessed on 30 June 2016), 2005.

[CAR 14] CARLBORG J., "Orange", available at: https://github.com/jacob-carlborg/orange, (accessed on 30 June 2016), 2014.

[CAS 15] CASTANEDA A., RAJSBAUM S., RAYNAL M., "Specifying concurrent problems: beyond linearizability and up to tasks", *Distributed Computing*, Springer, pp. 420–435, 2015.

[CHA 90] CHARRON-BOST B., "Concerning the size of clocks", *Proceedings of the LITP Spring School on Theoretical Computer Science: Semantics of Systems of Concurrent Processes*, pp. 176–184, Springer, 1990.

[COM 16] COMMUNITY T.D., "The D programming language", available at: http://dlang.org/, (accessed on 30 June 2016), 2016.

[COO 08] COOPER B.F., RAMAKRISHNAN R., SRIVASTAVA U. *et al.*, "PNUTS: Yahoo!'s hosted data serving platform", *Proceedings of the VLDB Endowment*, vol. 1, no. 2, pp. 1277–1288, VLDB Endowment, 2008.

[DEC 07] DECANDIA G., HASTORUN D., JAMPANI M. *et al.*, "Dynamo: amazon's highly available key-value store",*ACM SIGOPS Operating Systems Review*, vol. 41, pp. 205–220, ACM, 2007.

[DER 14] DERRICK J., DONGOL B., SCHELLHORN G. *et al.*, "Quiescent consistency: defining and verifying relaxed linearizability", *FM 2014: Formal Methods*, pp. 200–214, Springer, 2014.

[DOL 87] DOLEV D., DWORK C., STOCKMEYER L., "On the minimal synchronism needed for distributed consensus", *Journal of the ACM (JACM)*, vol. 34, no. 1, pp. 77–97, ACM, 1987.

[DUB 86] DUBOIS M., SCHEURICH C., BRIGGS F., "Memory access buffering in multiprocessors", *ACM SIGARCH Computer Architecture News*, vol. 14, no. 2, pp. 434–442, ACM, 1986.

[DUB 15] DUBOIS S., GUERRAOUI R., KUZNETSOV P. *et al.*, "The weakest failure detector for eventual consistency", *Proceedings of the 2015 ACM Symposium on Principles of Distributed Computing*, ACM, pp. 375–384, 2015.

[FAR 06] FARZAN A., MADHUSUDAN P., "Causal atomicity", *Computer Aided Verification*, Springer, pp. 315–328, 2006.

[FEK 01] FEKETE A., LYNCH N., SHVARTSMAN A., "Specifying and using a partitionable group communication service",*ACM Transactions on Computer Systems (TOCS)*, vol. 19, no. 2, pp. 171–216, ACM, 2001.

[FIS 85] FISCHER M.J., LYNCH N.A., PATERSON M.S., "Impossibility of distributed consensus with one faulty process", *Journal of the ACM (JACM)*, vol. 32, no. 2, pp. 374–382, ACM, 1985.

[FRI 15] FRIEDMAN R., RAYNAL M., TAÏANI F., "Fisheye consistency: keeping data in synch in a georeplicated world", *NETYS-3rd International Conference on NETwork sYStems*, pp. 246–262, 2015.

[GAM 00] GAMBHIRE P., KSHEMKALYANI A.D., "Reducing false causality in causal message ordering",*High Performance Computing-HiPC 2000*, pp. 61–72, Springer, 2000.

[GHA 90] GHARACHORLOO K., LENOSKI D., LAUDON J. *et al. Memory Consistency and Event Ordering in Scalable Shared-memory Multiprocessors*, vol. 18, ACM, 1990.

[GIL 02] GILBERT S., LYNCH N., "Brewer's conjecture and the feasibility of consistent, available, partition-tolerant web services", *ACM SIGACT News*, vol. 33, no. 2, pp. 51–59, ACM, 2002.

[GOO 91] GOODMAN J.R., Cache Consistency and Sequential Consistency, University of Wisconsin-Madison, Computer Sciences Department, 1991.

[GUE 08] GUERRAOUI R., KAPALKA M., "On the correctness of transactional memory", *Proceedings of the 13th ACM SIGPLAN Symposium on Principles and Practice of Parallel Programming*, ACM, pp. 175–184, 2008.

[HAA 07] HAAR S., JARD C., JOURDAN G.-V., "Testing input/output partial order automata", *Testing of Software and Communicating Systems*, pp. 171–185, Springer, 2007.

[HAD 88] HADZILACOS V., "A theory of reliability in database systems", *Journal of the ACM (JACM)*, vol. 35, no. 1, pp. 121–145, ACM, 1988.

[HAD 13] HADZILACOS V., TOUEG S., "Reliable broadcast and related problems", *Distributed Systems*, vol. 26, pp. 97–145, ACM, 2013.

[HAE 83] HAERDER T., REUTER A., "Principles of transaction-oriented database recovery", *ACM Computing Surveys (CSUR)*, vol. 15, no. 4, pp. 287–317, ACM, 1983.

[HÉL 06] HÉLARY J.-M., MILANI A., "About the efficiency of partial replication to implement distributed shared memory", *ICPP 2006 – International Conference on Parallel Processing, 2006*, IEEE, pp. 263–270, 2006.

[HER 90] HERLIHY M.P., WING J.M., "Linearizability: a correctness condition for concurrent objects", *ACM Transactions on Programming Languages and Systems (TOPLAS)*, vol. 12, no. 3, pp. 463–492, ACM, 1990.

[HER 91] HERLIHY M., "Wait-free synchronization", *ACM Transactions on Programming Languages and Systems (TOPLAS)*, vol. 13, no. 1, pp. 124–149, ACM, 1991.

[HER 93] HERLIHY M., MOSS J.E.B., "Transactional memory: architectural support for lock-free data structures", *Proceedings of the 20th Annual International Symposium on Computer Architecture*vol. 21, pp. 289–300, ACM, 1993.

[HUT 90] HUTTO P.W., AHAMAD M., "Slow memory: weakening consistency to enhance concurrency in distributed shared memories", *Proceedings of the 10th International Conference on Distributed Computing Systems, 1990*, IEEE, pp. 302–309, 1990.

[IMB 12] IMBS D., RAYNAL M., "Virtual world consistency: a condition for STM systems (with a versatile protocol with invisible read operations)", *Theoretical Computer Science*, vol. 444, pp. 113–127, Elsevier, 2012.

[JIM 08] JIMÉNEZ E., FERNÁNDEZ A., CHOLVI V., "A parametrized algorithm that implements sequential, causal, and cache memory consistencies", *Journal of Systems and Software*, vol. 81, no. 1, pp. 120–131, Elsevier, 2008.

[KAR 93] KARSENTY A., BEAUDOUIN-LAFON M., "An algorithm for distributed groupware applications", *Proceedings the 13th International Conference on Distributed Computing Systems*, IEEE, pp. 195–202, 1993.

[KEM 09] KEMME B., "One-copy-serializability", *Encyclopedia of Database Systems*, Springer, pp. 1947–1948, 2009.

[KLE 15] KLEPPMANN M., A critique of the CAP theorem, Report, University of Cambridge, 2015.

[KOB 10] KOBASHI H., YAMANE Y., MURATA M. *et al.*, "Eventually consistent transaction", *Proceedings of the 22nd IASTED International Conference on Parallel and Distributed Computing and Systems*, pp. 103–109, 2010.

[LAK 10] LAKSHMAN A., MALIK P., "Cassandra: a decentralized structured storage system", *ACM SIGOPS Operating Systems Review*, vol. 44, no. 2, pp. 35–40, 2010.

[LAM 78] LAMPORT L., "Time, clocks, and the ordering of events in a distributed system", *Communications of the ACM*, vol. 21, no. 7, pp. 558–565, 1978.

[LAM 79] LAMPORT L., "How to make a multiprocessor computer that correctly executes multiprocess programs", *IEEE Transactions on Computers*, vol. 100, no. 9, pp. 690–691, 1979.

[LAM 84] LAMPORT L., "Using time instead of timeout for fault-tolerant distributed systems", *ACM Transactions on Programming Languages and Systems (TOPLAS)*, vol. 6, no. 2, pp. 254–280, 1984.

[LAM 86] LAMPORT L., "On interprocess communication", *Distributed Computing*, vol. 1, no. 2, pp. 86–101, Springer, 1986.

[LI 00] LI D., ZHOU L., MUNTZ R.R., "A new paradigm of user intention preservation in realtime collaborative editing systems", *Proceedings of Seventh International Conference on Parallel and Distributed Systems, 2000*, IEEE, pp. 401–408, 2000.

[LIP 88] LIPTON R.J., SANDBERG J.S., PRAM: A Scalable Shared Memory, Princeton University, Department of Computer Science, 1988.

[LLO 11] LLOYD W., FREEDMAN M.J., KAMINSKY M. *et al.*, "Don't settle for eventual: scalable causal consistency for wide-area storage with COPS", *Proceedings of the Twenty-Third ACM Symposium on Operating Systems Principles*, ACM, pp. 401–416, 2011.

[MAH 11] MAHAJAN P., ALVISI L., DAHLIN M., Consistency, availability, and convergence, Report, University of Texas at Austin, 2011.

[MAU 15] MAUSSION D., Update consistency for data integrity in distributed data bases, Report, University of Nantes, 2015.

[MEA 55] MEALY G.H., "A method for synthesizing sequential circuits", *Bell System Technical Journal*, vol. 34, no. 5, pp. 1045–1079, Wiley Online Library, 1955.

[MIL 06] MILANI A., Causal consistency in static and dynamic distributed systems, PhD Thesis, "La Sapienza" Universita di Roma, 2006.

[MIS 86] MISRA J., "Axioms for memory access in asynchronous hardware systems", *ACM Transactions on Programming Languages and Systems (TOPLAS)*, vol. 8, no. 1, pp. 142–153, 1986.

[MOS 93] MOSBERGER D., "Memory consistency models", *ACM SIGOPS Operating Systems Review*, vol. 27, no. 1, pp. 18–26, ACM, 1993.

[MOS 15] MOSTEFAOUI A., PERRIN M., RUAS O., "CODS D library", available at: https://github.com/MatthieuPerrin/CODS, accessed on 30 July 2016, 2015.

[MUK 14] MUKUND M., SHENOY G., SURESH S., "Optimized or-sets without ordering constraints", *Distributed Computing and Networking*, vol. 8314, Springer, pp. 227–241, 2014.

[NÉD 13] NÉDELEC B., MOLLI P., MOSTEFAOUI A. *et al.*, "LSEQ: an adaptive structure for sequences in distributed collaborative editing", *Proceedings of the 2013 ACM Symposium on Document Engineering*, ACM, pp. 37–46, 2013.

[NEI 94] NEIGER G., "Set-linearizability", *Proceedings of the Thirteenth Annual ACM Symposium on Principles of Distributed Computing*, ACM, p. 396, 1994.

[OST 06] OSTER G., URSO P., MOLLI P. *et al.*, "Data consistency for P2P collaborative editing", *Proceedings of the 2006 20th Anniversary Conference on Computer Supported Cooperative Work*, ACM, pp. 259–268, 2006.

[PAP 79] PAPADIMITRIOU C.H., "The serializability of concurrent database updates", *Journal of the ACM (JACM)*, vol. 26, no. 4, pp. 631–653, ACM, 1979.

[PER 14] PERRIN M., MOSTÉFAOUI A., JARD C., "Brief announcement: update consistency in partitionable systems", *Proceedings of the 28th International Symposium on Distributed Computing*, Springer, p. 546, 2014.

[PER 15] PERRIN M., MOSTÉFAOUI A., JARD C., "Update consistency for wait-free concurrent objects", *Proceedings of the 29th IEEE International Parallel and Distributed Processing Symposium*, IEEE, pp. 219–228, 2015.

[PER 16] PERRIN M., PETROLIA M., MOSTEFAOUI A. *et al.*, "On composition and implementation of sequential consistency", *International Symposium on Distributed Computing*, Springer, pp. 284–297, 2016.

[PER 16a] PERRIN M., Spécification des objets partagés dans les systèmes sans-attente, PhD Thesis, University of Nantes, 2016.

[PER 16b] PERRIN M., MOSTÉFAOUI A., JARD C., "Causal consistency: beyond memory", *Proceedings of the 21st ACM SIGPLAN Symposium on Principles and Practice of Parallel Programming*, ACM, p. 26, 2016.

[PET 97] PETERSEN K., SPREITZER M.J., TERRY D.B. *et al.*, "Flexible update propagation for weakly consistent replication", *ACM SIGOPS Operating Systems Review*, vol. 31, pp. 288–301, 1997.

[PRE 09] PREGUICA N., MARQUES J.M., SHAPIRO M. *et al.*, "A commutative replicated data type for cooperative editing", *Distributed Computing Systems, 2009. 29th IEEE International Conference on ICDCS'09*, IEEE, pp. 395–403, 2009.

[RAY 91] RAYNAL M., SCHIPER A., TOUEG S., "The causal ordering abstraction and a simple way to implement it", *Information Processing Letters*, vol. 39, no. 6, pp. 343–350, 1991.

[RAY 96] RAYNAL M., SINGHAL M., "Logical time: capturing causality in distributed systems", *Computer*, vol. 29, no. 2, pp. 49–56, 1996.

[RAY 15] RAYNAL M., "Parallel Computing vs. Distributed Computing: A Great Confusion?(Position Paper)", *European Conference on Parallel Processing*, Springer, pp. 41–53, 2015.

[RUA 14] RUAS O., MOSTÉFAOUI A., PERRIN M., Weak consistency criteria: conception and implementation, Report, University of Nantes, 2014.

[SAI 05] SAITO Y., SHAPIRO M., "Optimistic replication", *ACM Computing Surveys (CSUR)*, vol. 37, no. 1, pp. 42–81, 2005.

[SCH 89] SCHIPER A., EGGLI J., SANDOZ A., "A new algorithm to implement causal ordering", *International Workshop on Distributed Algorithms*, Lecture Notes in Computer Science, vol. 392, Springer, pp. 219–232, 1989.

[SCH 90] SCHNEIDER F.B., "Implementing fault-tolerant services using the state machine approach: a tutorial", *ACM Computing Surveys (CSUR)*, vol. 22, no. 4, pp. 299–319, 1990.

[SER 10] SERAFINI M., DOBRE D., MAJUNTKE M. *et al.*, "Eventually linearizable shared objects", *Proceedings of the 29th ACM SIGACT-SIGOPS Symposium on Principles of Distributed Computing*, ACM, pp. 95–104, 2010.

[SEZ 15] SEZGIN A., Sequential Consistency and Concurrent Data Structures, Report, University of Cambridge, 2015.

[SHA 11a] SHAPIRO M., PREGUIÇA N., BAQUERO C. *et al.*, "Conflict-free replicated data types", *Stabilization, Safety, and Security of Distributed Systems*, Springer, pp. 386–400, 2011.

[SHA 11b] SHAPIRO M., PREGUIÇA N., BAQUERO C. *et al.*, A comprehensive study of convergent and commutative replicated data types, Report, UPMC, Paris, 2011.

[SHA 11c] SHAVIT N., "Data structures in the multicore age", *Communications of the ACM*, vol. 54, no. 3, pp. 76–84, 2011.

[SHE 97] SHENK E., "The consensus hierarchy is not robust", *Proceedings of the 16th Annual ACM Symposium on Principles of Distributed Computing*, p. 279, 1997.

[SOR 11] SORIN D.J., HILL M.D., WOOD D.A., "A primer on memory consistency and cache coherence", *Synthesis Lectures on Computer Architecture*, vol. 6, no. 3, pp. 1–212, 2011.

[SPA 99] SPACE SCIENCES LABORATORY U.B., "SETI@home", http://setiathome.ssl.berkeley.edu/, accessed 30 June 2016, 1999.

[SUN 98] SUN C., JIA X., ZHANG Y. *et al.*, "Achieving convergence, causality preservation, and intention preservation in real-time cooperative editing systems", *ACM Transactions on Computer-Human Interaction (TOCHI)*, vol. 5, no. 1, pp. 63–108, 1998.

[TEA 16] TEAM C.P.R., "FACEBOOK MaliciousChat", available at: http://blog.checkpoint.com/2016/06/07/facebook-maliciouschat/, accessed on 24 October 2016.

[TER 94] TERRY D.B., DEMERS A.J., PETERSEN K. *et al.*, "Session guarantees for weakly consistent replicated data", *Parallel and Distributed Information Systems, 1994, Proceedings of the Third International Conference on*, IEEE, pp. 140–149, 1994.

[TER 95] TERRY D.B., THEIMER M.M., PETERSEN K. *et al.*, "Managing update conflicts in Bayou, a weakly connected replicated storage system", *ACM SIGOPS Operating Systems Review*, vol. 29, pp. 172–182, 1995.

[TSE 15] TSENG L., BENZER A., VAIDYA N., Application-aware consistency: an application to social network, Report, University of Illinois at Urbana-Champaign, 2015.

[VIT 03] VITENBERG R., FRIEDMAN R., "On the locality of consistency conditions", *International Symposium on Distributed Computing*, vol. 2848, Springer, pp. 92–105, 2003.

[VOG 09] VOGELS W., "Eventually consistent", *Communications of the ACM*, vol. 52, no. 1, pp. 40–44, 2009.

[WEI 89] WEIHL W.E., "Local atomicity properties: modular concurrency control for abstract data types", *ACM Transactions on Programming Languages and Systems (TOPLAS)*, vol. 11, no. 2, pp. 249–282, 1989.

[WEI 09] WEISS S., URSO P., MOLLI P., "Logoot: a scalable optimistic replication algorithm for collaborative editing on P2P networks", *Distributed Computing Systems, 2009. 29th IEEE International Conference on ICDCS'09*, IEEE, pp. 404–412, 2009.

[WUU 86] WUU G.T., BERNSTEIN A.J., "Efficient solutions to the replicated log and dictionary problems", *Operating Systems Review*, vol. 20, no. 1, pp. 57–66, 1986.

[XIE 14] XIE C., SU C., KAPRITSOS M. *et al.*, "Salt: combining ACID and BASE in a distributed database", *Proceedings of the 11th USENIX Conference on Operating Systems Design and Implementation*, vol. 14, pp. 495–509, Berkeley, CA, 2014.

Index

H, I, K, L, M

O, P, Q, R, S

T, U, V, W